Soil Change Matters 2014

IOP Conference Series: Earth and Environmental Science
Volume 25

Bendigo, Victoria, Australia
24 - 27 March 2014

ISBN: 978-1-5108-0991-8
ISSN: 1755-1307

Printed from e-media with permission by:

Curran Associates, Inc.
57 Morehouse Lane
Red Hook, NY 12571

Some format issues inherent in the e-media version may also appear in this print version.

Printed by Curran Associates, Inc. (2015)

For permission requests, please contact the Institute of Physics at the address below.

Institute of Physics
Dirac House, Temple Back
Bristol BS1 6BE UK

Phone: 44 1 17 929 7481
Fax: 44 1 17 920 0979

techtracking@iop.org

Additional copies of this publication are available from:

Curran Associates, Inc.
57 Morehouse Lane
Red Hook, NY 12571 USA
Phone: 845-758-0400
Fax: 845-758-2634
Email: curran@proceedings.com
Web: www.proceedings.com

Table of contents

Volume 25
Soil Change Matters 2014
24–27 March 2014, Bendigo, Victoria, Australia

Accepted papers received: 11 June 2015
Published online: 17 July 2015

Preface

Papers

IOP Conf. Series: Earth and Environmental Science 25 (2015) 011001 doi:10.1088/1755-1315/25/1/011001

Preface

The opinions expressed and arguments employed in this publication are the sole responsibility of the authors and do not necessarily reflect those of the OECD or of the governments of its Member countries.

The Workshop was sponsored by the OECD Co-operative Research Programme on Biological Resource Management for Sustainable Agricultural Systems, whose financial support made it possible for eight of the invited speakers to participate in the Workshop.

Principal Sponsor

OECD
BETTER POLICIES FOR BETTER LIVES

Gold sponsor

Australian Government
Department of Agriculture

Silver sponsor

CSIRO

Bronze sponsors

IPNI
INTERNATIONAL PLANT NUTRITION INSTITUTE

THE UNIVERSITY OF SYDNEY

une University of New England
School of Environmental and Rural Science

NORTH CENTRAL
Catchment Management Authority
Connecting Rivers, Landscapes, People

Supporters

We would like to thank the Organising Committee, the Scientific Committee and the financial support from the conference sponsors and funding from the Government of Victoria that allowed the success of the Soil Change Matters Workshop.

Organising Committee (Department of Economic Development, Jobs, Transport and Resources unless otherwise noted):

- Richard MacEwan (Convenor)
- Jennifer Alexander
- Helaina Black (James Hutton Institute, UK)
- Doug Crawford
- Phil Dyson (North Central CMA)
- Jane Fisher
- Gemma Heemskerk
- Jonathan Hopley
- Pauline Mele
- Rebecca Mitchell
- David Rees
- Dugal Wallace
- Dale Webster

Scientific Committee (Department of Economic Development, Jobs, Transport and Resources unless otherwise noted):

- Mr Richard MacEwan
- Dr Dominique Arrouays (National Institute of Agronomic Research, France)
- Helaina Black (James Hutton Institute, UK)
- Mr Doug Crawford
- Dr Ben Marchant (Geoscience, UK)
- Dr Pauline Mele
- Dr Budiman Minasny (University of Sydney, NSW)
- Professor Dan Richter (Duke University, USA)
- Mr Nathan Robinson

Thanks are given to the authors and to the anonymous referees for the papers included here.

Soil Change Matters Workshop 24-27 March 2014 - Photographs

Figure 1 Forty delegates attended the pre-workshop fieldtrip. The photograph is at one of the four sites visited - a soil pit at Balgownie Estate vineyard, Bendigo. Photographer: Ron Aggs

Figure 2 OECD sponsored delegates (L to R): Prof Johan Bouma (Netherlands), Prof Dan Richter (USA), Maxine Levin (USA), Mary-Vaughn Johnson (USA), Daniela Sauer (Germany), John Sadler (OECD representative), Helaina Black (Scotland, DEPI visiting fellow), Ben Marchant (UK), Allan Lilly (UK), Pierre Roudier (NZ). Photographer: Ron Aggs

Figure 3 Workshop logistics team: Richard MacEwan, Jennifer Alexander, Rebecca Mitchell. Photographer: Dale Webster

Figure 4 Members of the National Soil Research Development and Extension Strategy Reference Group (L to R). Prof Alex McBratney, Prof Iain Young, Michele Barson, Georgina Kelly, Michael Crawford, Caroline Wardrop, Noel Schoknecht. The Strategy was launched at the workshop. Photographer: Ron Aggs

Figure 5 Luca Montanarella (EU), chair of the FAO's Intergovernmental Technical Panel on soils (Global Soil Partnership) addresses the workshop. Photographer: Erin Jancauskas

Figure 6 Helaina Black, Andrea Koch, Linda Bennison and David Cummings fill a tablecloth at the soil policy 'world café'. Photographer: Erin Jancauskas

Figure 7 World café conversations. Foreground table (L to R): Chuck Rice, Pauline Mele, Johan Bouma and Neil McKenzie. Photographer: Erin Jancauskas

Figure 8 Songs of the soil and farming formed part of the evening soil show "Soil Matters". (Richard MacEwan & Chris Jacobs). Photographer: Helaina Black

Figure 9 Stringband 'Buffalo Nickel', provided music for the workshop buffet dinner. Conference convenor, Richard could not resist joining in on banjo for a couple of numbers. Photographer: Ron Aggs

IOP Conf. Series: Earth and Environmental Science **25** (2015) 011002 doi:10.1088/1755-1315/25/1/011002

Figure 10 Matt Kitching (right) demonstrated handheld mid infrared spectroscopy during the field trip. (L to R) Doug Crawford, Sharon Aarons and Rob Clark. Photographer: Ron Aggs

Figure 11 John Leys explains the national Dustwatch and wind erosion monitoring work at the southernmost dust monitoring station in Australia. Photographer: Ron Aggs

Figure 12 DEPI Department Secretary, Adam Fennessy welcomed delegates and chaired the opening session (Day 1) Photographer: Ron Aggs

Figure 13 Major General Michael Jeffery, The Australian Advocate for Soil Health gave the opening address at the Soil Matters Symposium (Day 1) Photographer: Ron Aggs

Figure 14 The OECD Cooperative Research Program representative, John Sadler gets a grip on documenting the group discussions during the Soil Policy Workshop on Wednesday afternoon (Day 2) Photographer: Richard MacEwan

Figure 15 The final plenary session and Q&A with (L to R) Luca Montanarella, Michele Barson, John Leys and Fiona Robertson. Photographer: Richard MacEwan

Figure 16 Soil Science Australia Executive Officer, Linda Bennison, with Richard MacEwan (past president) and Richard Doyle (president). Photographer: Ron Aggs

Figure 17 Pierre Roudier gave the final keynote presentation of the workshop, speaking about the 'data deluge' and challenges for soil scientists. Photographer: Ron Aggs

Figure 18 Johan Bouma summed up the workshop, thanking the organisers and presenting Richard MacEwan, the conference convenor, with a fine spade. Photographer; Helaina Black

Figure 19 The workshop closing remarks were from Dr John Sadler the OECD CRP representative who said the workshop 'exceeded expectations, particularly for the engagement of science with policy'. Photographer: Helaina Black

Figure 20 Final picture of the delegates present for a group photograph in the theatre at the end of the workshop. Photographer: Ron Aggs

IOP Conf. Series: Earth and Environmental Science **25** (2015) 011002 doi:10.1088/1755-1315/25/1/011002

Figure 21 Graphic recording of the workshop provided by Erin Jancauskas is a visual legacy in a 15 metre long cartoon representation of some of the keynote talks and points. Three portions shown here. Photographer: Erin Jancauskas

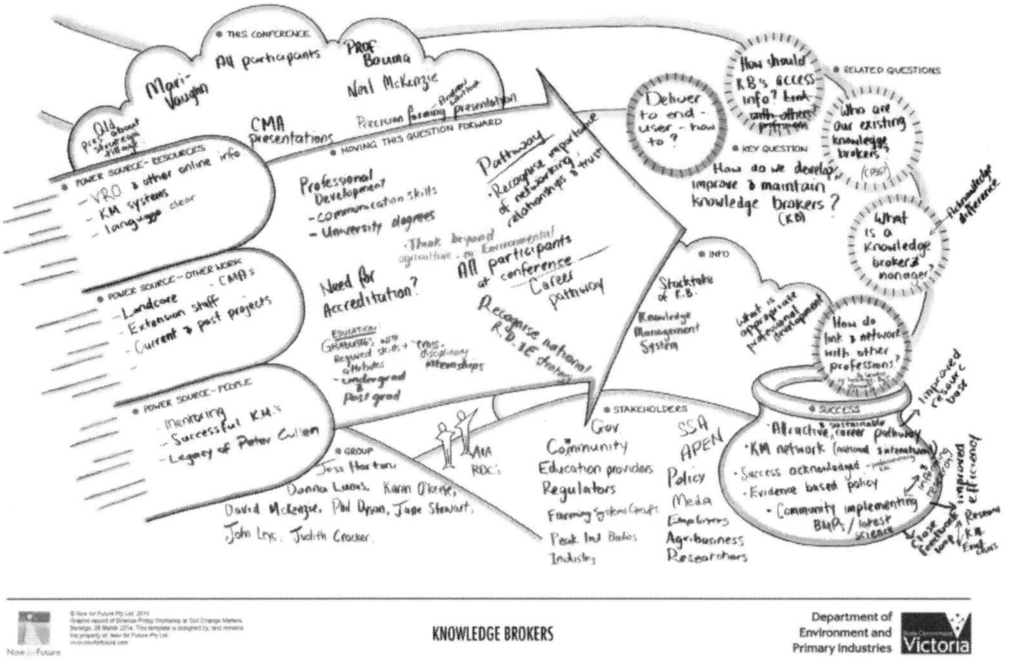

Figure 22 One of the five graphic outputs produced by discussion groups during the science and policy workshop. Photographer: Erin Jancauskas

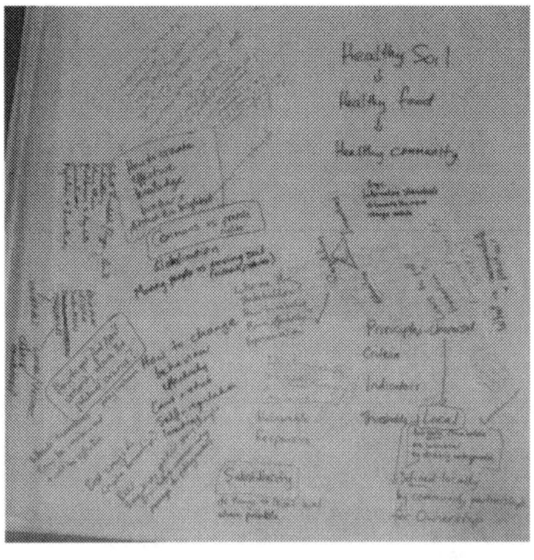

Figure 23 One of the 13 'tablecloth' discussion records from the soil policy workshop world café. Photographer: Grant Boyle

Figure 24 'Now For Future' facilitators Jenni Dunn and Erin Jancauskas discuss progress during a break in the science and policy workshop. Photographer: Ron Aggs

Soil Change Matters International Workshop – Overview

Richard MacEwan and Jennifer Alexander

Agriculture Research, Department of Economic Development, Jobs, Transport and Resources, Victoria
Box 3100 Bendigo Delivery Centre, Victoria Australia 3554.

E-mail richard.macewan@ecodev.vic.gov.au

Abstract. Soils change in response to land use, land management and climate. Understanding the mechanisms and rates of change in fundamental soil properties, their extent across landscapes, and the drivers of change, is critical for management of soil and land to ensure enduring productivity and the provision of ecosystem services. A reliable evidence base for soil change can be created through research into fundamental soil properties and processes, which coupled with monitoring, can be used to develop models to support management and policy decisions. Soil scientists, land managers and policy makers occupy different intellectual domains and each can contribute to a soil change dialogue focussed on providing the best science for managers, policy and decision-making. Clarity is needed concerning the knowledge needs of policy makers and the latter need to understand what is achievable by science and at what cost. However, communication is often limited by the discipline-specific language used by each group. Facilitated communication would improve the impact of science on policy by enabling scientists to better understand policy maker needs and enabling policy makers to better understand scientific data interpretations, limitations, and gaps. The international workshop 'Soil Change Matters' in Bendigo, Australia 24-27 March 2014, brought together 200 delegates across these intellectual domains to talk to each other, share conclusions, questions and data, and to learn about scenarios for soil change and for soil policies.

1. Introduction

The International Workshop 'Soil Change Matters' in Bendigo, Australia, 24-27 March 2014, was convened to explore and discuss changes to soil in the Anthropocene (the era of human influence on the environment) – specifically, how our use of land and soil has affected soil quality and what policy responses are needed to ensure sustainability of the soil resource. The workshop brought together scientists, policy makers, land managers and other stakeholders for a forum in which over 70 separate presentations were made, ranging from the newest perspectives on soil biology to international policies for soil protection.

Two hundred delegates attended, representing 11 countries (Australia, USA, UK, New Zealand, France, Netherlands, Germany, Russia, Slovakia, Italy and Canada). Of the 200 delegates, 45% were identifiable as actively engaged in research (14.5% in Academia, 30.5% in Government), 20% as engaged in extension services or land management (including a few landholders), 15% engaged in policy at some level (including international, national, state, and catchment authorities), 9.5% were consultants, 5% students, and the remainder (6%) communications, journalism and sales.

A field trip prior to the workshop provided an opportunity for 40 of the delegates to get to know each other, visit some farm sites, interact with farmers and local government officials to discuss the land management and production issues relevant to the soils and sites visited. Spaces on this trip were preferentially allocated to all of the invited and sponsored speakers and to representatives from across the groups mentioned above.

The format for the workshop itself allowed over 40% of the time for discussion by limiting 50% of the presentations to 5 minute synopses. In an intensive 2.5 days, excluding breaks, 14 hours were scheduled for talks and 10 hours for questions and discussion, in plenary and parallel sessions. Seventeen of the speakers were invited to provide concise keynote and thematic talks (15-25 minutes each).

Twenty-three presenters have provided papers for review and publication in this volume of the IOP Earth and Environmental Science series, and eleven presenters have contributed papers to a special issue of Soil Research. All of the abstracts and some information for the workshop can be viewed online at www.soilmatters.org.

2. Workshop objectives

The workshop objective was to gain a better understanding of current trends and critical changes in soil to determine future research, development and policy priorities. Specifically, the workshop goals were to:

1) provide a forum to present scientific research related to understanding changes happening to soil under current land uses and management;
2) facilitate a workshop to clarify the role of policy in soil security and define the scientific capacity to provide technical understanding and guidance to policy;
3) stimulate collaboration across research organisations and four International Union of Soil Science working groups (soil monitoring; proximal soil sensing; global soil change; digital soil mapping);
4) provide a forum for the latest developments in soil science to engage a variety of stakeholders, including other scientists, land managers and the general public; and
5) showcase policy development and priorities for investment at a national level through the National Soils Research, Development and Extension (RD&E) strategy, and explore the possible synergies with international initiatives such as the United Nations Food and Agriculture Organization's (FAO) Global Soil Partnership.

3. Papers presented by the OECD sponsored speakers

The Organisation for Economic Co-operation and Development (OECD) Cooperative Research Program (CRP) funded eight international invited speakers. The papers from each of the OECD supported delegates are included in this volume and are as follows. With their co-authors, Ms Maxine Levin ('Delivering solutions to questions regarding soil change—examples from USDA and the National Cooperative Soil Survey'), and Dr Mari-Vaughn Johnson ('The Conservation Effects Assessment Project (CEAP): a national scale natural resources and conservation needs assessment and decision support tool') give accounts of the USDA's contribution to understanding and monitoring soil change, as well as providing some insights into how scientific information informs soil and natural resource policy in the United States.

Dr Daniela Sauer (Germany) and Professor Daniel Richter (USA) provide two pedological perspectives for the temporal scales of soil change, firstly that of landscape evolution and soil research spanning soil changes over millennia ('Soil development over millennial timescales – a comparison of soil chronosequences of different climates and lithologies') and secondly, the importance of recognising the Anthropocene era as a period of accelerated soil change for which human activity is responsible ('Soil in the Anthropocene'). Professor Johan Bouma ('Policy and effective action for soil security: a need for reframing the soil story') makes an appeal for examples of soil scientists engaging in inter- and transdisciplinary action for real world problem solving. He calls this 'reframing the soil narrative' and provides an example from the Netherlands where a program, with soil scientists as knowledge brokers at the hub, successfully addressed five major environmental issues (food security, water and energy availability, climate change and biodiversity loss) in an effort that resulted in a change of policy for dairy farmers' use of fertiliser and manure.

Dr Ben Marchant and Dr Allan Lilly present two case studies for monitoring soil change using legacy data and samples. The paper by Marchant et al. ('What can legacy datasets tell us about soil quality trends? Soil acidity in Victoria') presents an innovative geo-statistical approach to the use of large datasets, time-stamped but geographically generalised, from laboratory services to create a statewide map of changes in soil surface pH over time. Lilly and Chapman describe the approach used to resample Scottish soils and re-analyse archived soil samples, taken 20-30 years previously, to determine whether soil organic carbon (SOC) has changed ('Assessing changes in carbon stocks of Scottish soils: lessons learnt'). Their sampling and analytical methodology is thorough, including the use of Fourier transform infrared spectroscopy (FTIR) to confirm comparable samples from the locations for the earlier and recent dates. Their conclusion is that there is no detectable loss of SOC from Scottish soils in that period. Dr Pierre Roudier ('The rise of information science: a changing landscape for soil science') introduces the notion of a 'data deluge' and addresses major questions concerning the implications for large volumes of data generated in soil science and the potential for their analysis.

4. Structure of this volume of IOP Earth and Environmental Sciences, including other invited and contributed papers

The workshop was divided into sessions that focussed on different aspects of soil change including fundamental soil properties, monitoring, mapping and modelling soil change, land use and land management, policy and soil change, and the issues around data and data management. The papers included in this volume represent only 30% of the total presentations given at the workshop and have been grouped according to their major emphasis, although they may cover more than one of these aspects. Other workshop presentations, not included here, that help to complete a picture of the scope of the workshop are mentioned below and, where possible, citations of their publication in other journals are included.

4.1 Integrating soil science into policy

A primary objective for the workshop was to discuss policies supporting soil management. Policy related papers accepted for this volume include Luca Montanarella's explanation of the Global Soil Partnership (GSP), Drs Jane Fisher and Michael Crawford's perspective on a state government's approach to soil policy ('The policy and science of soil change – a Victorian perspective'), Professor Robert White and Dr Brian Davidson's analysis of the cost effectiveness of C-sequestration in Australian agricultural systems, and Professor Johan Bouma's paper on reframing the soil narrative. 'Soil Security' is a term that has recently emerged and is particularly relevant to policy. A paper by Dr Damien Field et al. explained the term, its breadth and inclusions, and is published in the journal Geoderma [1]. The paper by Mr Ron Aggs ('Soil news – the soil carbon and climate policy journey in Australia and the role of different media'), tackles the important topic of communication for soil science and the issues affecting success of policy development and acceptance, using, as an example, the Carbon Farming Initiative in Australia. Aggs points out the roles and limitations for different media in getting messages across to the voting populace.

4.2 Fundamental soil properties and rates of change

The fundamental science of soil properties and soil change is found in the invited papers by Drs Helaina Black and Pauline Mele ('Advances in Soil Biology: What does this mean for assessing soil change?'), Professor Mike McLaughlin ('The Soil and its Chemistry - Critical Futures'), and Dr Brian Murphy's review of relationships between organic matter and soil properties ('Key soil functional properties affected by soil organic matter – evidence from published literature').

Understanding how individual soil properties change over time, rates of soil formation and the drivers of change is essential if we are to model the impacts of land use practices and their effects on soil sustainability. The workshop heard three presentations that addressed this topic. The papers by Drs Sauer and Richter have already been referred to above. Sauer and her co-authors show how the study of chronosequences of soil, aged from a few hundred to several hundred thousand years, can reveal rates of soil forming processes, and Richter et al. focus on the important anthropogenic influences. A third complementary paper was presented by Dr Uta Stockmann where she and her fellow researchers have developed algorithms for modelling soil formation rates over millennia. Some of this work is published in the journal Geoderma and can be referred to there [3].

4.3 Monitoring soil change

In order to understand, quantify and model soil change in the Anthropocene (the era of human influence on the environment) there is a fundamental need for primary data on soil properties, in different land use, landscape and climate contexts. The collection of such data occurs primarily in specific research projects since examples of long-term soil condition monitoring are rare; however some examples were presented at the workshop. Professor Josef Kobza ('Permanent soil monitoring system as a basic tool for protection of soils and sustainable land use in Slovakia') describes the 20 year old network of 318 sites on agricultural land in Slovakia and how the data collected are included in broader EU data sets. An important aspect of the monitoring is that it provides the evidence base for development of policy and legislation for land use practices and soil protection. The CEAP project in the United States, described by Dr Mari-Vaughn Johnson, is a more recent monitoring and modelling exercise, in which the baseline data were collected ten years ago and follow-up data collection is scheduled for 2015-16. The data are analysed and applied to process-based modelling exercises across 10 river basins, the results of which are used to interpret the adequacy of current conservation and to determine areas of land in need of additional conservation practices. The CEAP project is a clear example and demonstration of the important contribution that science-based monitoring and modelling can make to the development and support of soil conservation policies.

Remote sensing continues to develop and offer opportunities for monitoring. The paper by Drs Kathryn Sheffield and Liz Morse-McNabb ('Using satellite imagery to assess trends in soil and crop productivity across landscapes') presents current research into application of MODIS satellite to monitor crop productivity across the landscape, and a legacy of Landsat data over 40 years to look for trends and patterns. Crops integrate the properties of the soil with the practices of the land manager and the seasonal climate so there is potential for the satellite NDVI to be used as an indicator of soil condition if the other factors are known and sufficient soil ground truth exists or is collected. At the paddock scale, the paper by Dale Boyd ('Soil moisture monitoring for crop management') describes a soil moisture monitoring network created for farmers in Victoria as an input into the seasonal decision making for strategic fertilizer application.

4.4 Land use and soil change

A major focus of soil research currently is SOC driven by the climate change debate and the desire to engineer land use practices that can remove more CO_2 from the atmosphere, sequestering the carbon in the soil. Reflecting this interest, 20 of the papers offered to the workshop had SOC as their focus. This includes papers by Professor Robert White and Dr Brian Davidson, referred to in 4.1, Dr Brian Murphy's paper referred to in 4.2, and Lilly and Chapman's paper on SOC in Scottish Soils. Dr Garry O'Leary and his co-authors ('Modelling soil carbon in agricultural systems: a way to widen the experimental space') stress the need to develop and apply process models for SOC (rather than purely statistical models) and to link biophysical and economic (agricultural productivity) models in order to provide more realistic or practical scenarios for C sequestration. Their point is that land use practices will remain unchanged unless there are productivity gains, not merely C sequestration, and better modelling is needed to predict the outcomes for land use change scenarios.

Several papers presented surveys, modelling and analyses of SOC under particular land uses: Susan Orgill's presentation ('Soil carbon under perennial pastures; benchmarking the influence of pasture age and management') suggests a relationship between SOC land use management and parent material and emphasises the importance of pasture species and fertiliser additions in maintaining or increasing SOC in pasture, but there are many research questions to be answered on this topic. Dr Sharon Aarons and co-authors ('Can soil change be assessed for the Victorian dairy industry?') describe the use of legacy data, firstly to quantify the areas of different soil orders under dairy management in Victoria, Australia, and secondly to assess if soil properties for electrical conductivity, pH, potassium and phosphorus had changed under dairy management. They caution that while legacy data can be useful for assessing soil change, researchers need to be mindful of 'fit for purpose' sampling when interpreting data.

4.5 Data, information and knowledge management

Soil samples, field descriptions, laboratory analyses, aerial photography, and imagery from satellite sensors provide the primary data for understanding soil, landscape and land use. These samples and data have lasting value, particularly to answer questions regarding soil change. Many presentations and discussions at the workshop referred to this and there are principles in policy that mirror this concern– in pillar four of the GSP to 'Enhance the quantity and quality of soil data and information: data collection (generation), analysis, validation, reporting, monitoring and integration with other disciplines'[4] – and in the goals of the (Australian) RDE strategy for soils to 'Improve quality, availability and access to soil data and information'[2].

Legacy data from soil surveys, field experiments and monitoring projects are invaluable for applications such as digital soil mapping (DSM) and interpreted products for planning, assessing land capability or, in conjunction with ancillary data on land use practices, for assessing likely and actual changes in soil quality. The paper presented by Ms Maxine Levin describes how use of the National Cooperative Soil Survey (NCSS) data has evolved in the USA and is increasingly being used to support a variety of end users and policy objectives for land use and ecological monitoring. The work reported in the paper by Marchant et al (this volume) referred to above used legacy data from 75,000 routine soil tests on samples submitted by farmers over two decades (1973-93) to map change in surface soil pH.

In the context for understanding soil change, we need to be aware that data on soil properties can be unlocked from a sample at any time if sufficient sample has been carefully archived. The value of soil archives is increasingly being recognised, particularly as methods of analysis are improved over time to achieve greater precision and accuracy. A contemporary sample can be more reliably compared with an archived sample if both are subjected to identical analytical methods and, an archived sample can be retrieved and analysed to yield data on properties not previously measured. Linda Karsies and Peter Wilson ('Soil Archives: supporting Research into Soil Changes') explain the process and value of the national soil archive managed by CSIRO in Canberra and provide an example of how samples in the Victorian soil archive were used to correct SOC measurements recorded over a period during which the laboratory had applied changes to the Walkley-Black method. Without the archived sample to re-analyse, changes in SOC would have been predicted – a similar example for the value of archiving is given for the Scottish study presented by Allan Lilly in these proceedings. Currently, soil archives around the world are proving their worth through the acquisition of spectroscopy, such as mid infra-red, to create predictive calibration sets for a suite of soil properties.

Making information available to others is a critical function of all research, and although the traditional practice is to publish research in peer-reviewed journals, these published papers only represent a fraction of the data generated during the research and only reach a fraction of the potential audience of stakeholders. In the case of soil survey, publication online of primary data sets as well as maps is now expected. Kelly Bryant and co-authors ('Accessing Queensland's soil information – an open data revolution!') describe one State Government's approach to data and information delivery via the web. Pierre Roudier and co-authors' keynote paper was the final for the workshop and provides a clear overview of the changes in information technology and the challenges and opportunities for soil scientists. Roudier makes an important point in this paper, saying, '...the challenges raised by the "data deluge" are not so much the amount of data per se, but the novel analysis to produce knowledge and help insightful decisions'.

5. Soil Change Matters – The Policy Dimension

Policy for soil is currently a conspicuous matter in national and international agenda. The recent formation in 2012 of the 'Global Soil Partnership' [4] and the Australian RDE strategy for soils [2], as well as the long-standing regional catchment bodies in Australia, provide a platform for integration of policy from global to local geographical scales. The Australian government's Advocate for Soil Health, Major General Michael Jeffery, provided a considered perspective on the challenges that are being faced for land management in Australia ending with a memorable quotation 'To save the planet, we have got to save the soil'. International delegates were particularly impressed by the fact that Australia has an advocate for soils, who is not a soil scientist, but, as an ex-Governor General for Australia, is well connected to Commonwealth and State politicians and is operating at a high level in government. The opening address from the Advocate, the presentations by Montanarella, Fisher and Crawford, and Bouma (this volume), and the launch of the RDE soil strategy set the atmosphere for some informed conversation and exploration of issues and solutions.

A half day facilitated soil science-soil policy workshop was attended by 45 of the delegates representing a range of interests from farming, planning, education, research, extension and policy.

Participants were invited to 'Imagine a future where:

- Our soil is secure – scientists and policy makers have made this a success;
- Governments provide long-term vision, guidance and assistance;
- Research is targeted and timely;
- Scientists and policy makers work as a team; and
- Researchers provide policy makers with exactly what they need.'

The task was then to determine what questions need to be addressed in order to achieve this future. Two rounds of discussions were held at tables of 4 and recorded on paper tablecloths. The discussions were then summarised into 5 focus areas for discussion: 'Our Message', 'Knowledge Brokers', 'Evidence to Drive Policy', 'What Does Success Look Like', and 'Connection Between Community and Soil'. Graphic templates, designed and prepared prior to the workshop in consultation between the conference executive and the contracted facilitators, were used to provide structure for recording the focussed discussions. The completed template for the 'Knowledge Brokers' focus group and an example of a tablecloth record from the world café can be seen in the workshop photographs (figures 22 and 23). The exercise was useful in that the many conversations held on that afternoon served to inform and excite the participants. To what extent this activity will engender action and stimulate change in any science or policy arena is indeterminate, however the primary objective - to get stakeholders from different domains to engage in these important questions - was achieved.

6. Conclusion

The workshop was described as highly successful by Dr John Sadler, OECD's CRP representative, particularly for the role it performed in encouraging a debate to clarify the science policy dimensions, highlighted by a full afternoon of 'world café' table-based and group discussions. A highlight for Australian policy at the workshop was the launch of the National Soil Research, Development and Extension Strategy [2] and an outline of its goals and strategies given by Dr Michele Barson from the Commonwealth Government's Department of Agriculture. The continuance of Major General Michael Jeffery as the advocate for soil health provides a strong presence in the policy arena for Australia.

7. Acknowledgments

The organisation of the Soil Change Matters workshop was funded by the Government of Victoria through the 'Understanding Soil and Farming Systems' project and benefited from the generous support of a number of sponsors: Principal sponsor: OECD Co-operative Research Programme on Biological Resource Management for Sustainable Agricultural Systems; Gold sponsor: Department of

Agriculture, Australian Government; Silver sponsor: CSIRO; Bronze sponsor: International Plant Nutrition Institute, University of Sydney, University of New England School of Environmental and Rural Science, North Central Catchment Management Authority; Supporters: International Union of Soil Sciences, British Society of Soil Science, Soil Science Australia, Telstra, Wiley, City of Greater Bendigo. Thanks to Professor Robert White and Dr Mari-Vaughn Johnson for their reviews that improved this paper, and to the authors and referees of the papers included here.

References

[1] McBratney, A., Field, D.J., Koch, A., 2014. The dimensions of soil security. Geoderma 213, 203-213.

[2] Commonwealth of Australia. 2014. Securing Australia's Soil for Profitable Industries and Healthy Landscapes. The National Soil Research, Development and Extension Strategy. http://www.agriculture.gov.au/ag-farm-food/natural-resources/soils/national_soil_rd_and_e_strategy

[3] Stockmann, U., Minasny, B., McBratney, A.B., 2014. How fast does soil grow? Geoderma 216, 48-61.

[4] http://www.fao.org/globalsoilpartnership/en/

Peer review statement

All papers published in this volume of *IOP Conference Series: Earth and Environmental Science* have been peer reviewed through processes administered by the proceedings Editors. Reviews were conducted by expert referees to the professional and scientific standards expected of a proceedings journal published by IOP Publishing.

The Global Soil Partnership

Luca Montanarella

European Commission - DG JRC, Via E. Fermi, 2749, I-21027 Ispra (VA), ITALY

E-mail: luca.montanarella@jrc.ec.europa.eu

Abstract. The Global Soil Partnership (GSP) has been established, following an intensive preparatory work of the Food and Agriculture Organization of the United Nations (FAO) in collaboration with the European Commission (EC), as a voluntary partnership coordinated by the FAO in September 2011 [1]. The GSP is open to all interested stakeholders: Governments (FAO Member States), Universities, Research Organizations, Civil Society Organizations, Industry and private companies. It is a voluntary partnership aiming towards providing a platform for active engagement in sustainable soil management and soil protection at all scales: local, national, regional and global. As a "coalition of the willing" towards soil protection, it attempts to make progress in reversing soil degradation with those partners that have a genuine will of protecting soils for our future generations. It openly aims towards creating an enabling environment, despite the resistance of a minority of national governments, for effective soil protection in the large majority of the countries that are genuinely concerned about the rapid depletion of their limited soil resources.

1. The Global Soil Partnership in Action

The Global Soil Partnership (GSP) partners are the owners of the GSP. Therefore the GSP Plenary assembly is the main decision making body of the partnership. It meets annually and is tasked with all relevant decisions for the GSP, like the approval of the various plan of action, the nomination of the Intergovernmental Technical Panel on Soils (ITPS) and the coordination of the Regional Partnerships. The GSP is supported by a Secretariat hosted by FAO (Figure 1).

The GSP is active in 5 main pillars of action:

1- Promote sustainable management of soil resources for soil protection, conservation and sustainable productivity

2- Encourage investment, technical cooperation, policy, education awareness and extension in soil

3- Promote targeted soil research and development focusing on identified gaps and priorities and synergies with related productive, environmental and social development actions

4- Enhance the quantity and quality of soil data and information: data collection (generation), analysis, validation, reporting, monitoring and integration with other disciplines

5- Harmonization of methods, measurements and indicators for the sustainable management and protection of soil resources

For each of these pillars detailed plans of action are developed with the help of specific working groups assembling the most relevant stakeholders and experts in relation to each of the relevant subjects. The highest scientific and technical level is guaranteed by the Intergovernmental Technical Panel on Soils (ITPS), as a core component of the GSP.

Figure 1. Organizational structure of the GSP.

2. The Intergovernmental Technical Panel on Soils (ITPS)

The ITPS is a scientific panel of 27 high level scientists selected according to a regional balance as follows:

- five from Africa
- five from Asia
- five from Europe
- five from Latin America and the Caribbean
- three from Near East
- two from North America
- two from South West Pacific

Given its intergovernmental nature, panel members are nominated by the FAO Member Countries on the basis of a shortlist prepared by the GSP Secretariat (FAO). The mandate of the panel members is of two years and can be renewed for an additional term of two years (Figure 2). Panel members are acting as independent experts and are not bound to respond to their respective governments. They provide the necessary scientific and technical advice to the GSP as well as to other UN bodies requiring soil related expertise, like the UNFCCC, the CBD and the UNCCD. As a science policy interface, the ITPS operates as similar bodies for other policy areas, like IPCC for climate change, IPBES for biodiversity and the new SPI for desertification.

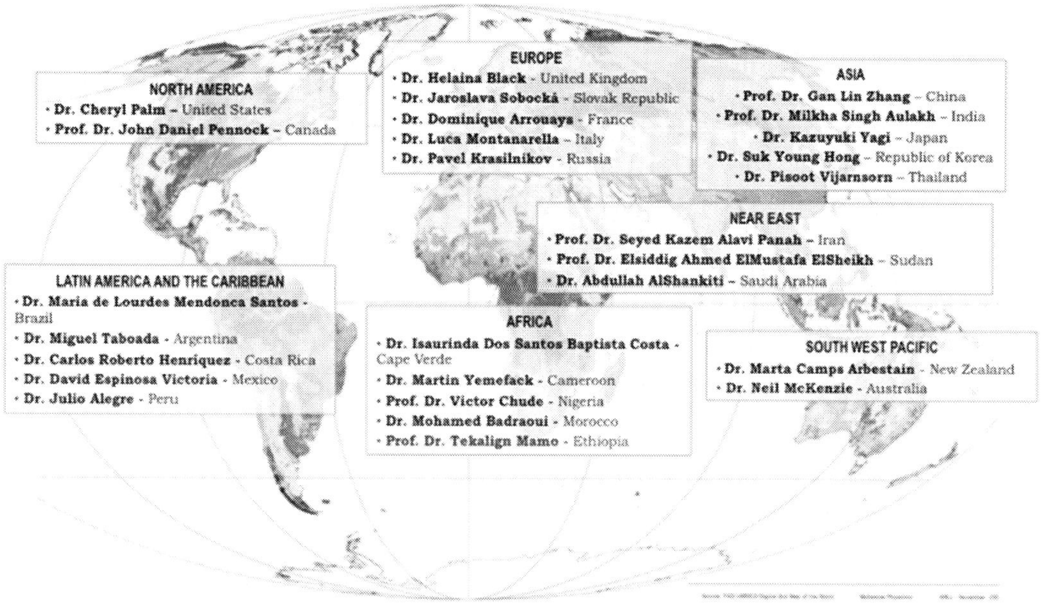

Figure 2. Members of the 1st Intergovernmental Technical Panel on Soils (ITPS) 2013-2015.

The ITPS, at its first meeting in July 2013 has identified its main activities and planned deliverables for its mandate of two years:

1. Development of the background scientific and technical evidence towards the inclusion of soils within the sustainable development agenda post-2015, especially in the frame of the outcome of the Rio+20 sustainable development conference. Particularly the development of a soil related sustainable development goal (SDG) is addressed by this scientific and technical support activity.
2. Revision and update of the World Soil Charter. The FAO World Soil Charter was adopted in 1981 by the 21st session of the FAO Conference and establishes a set of principles for the optimum use of the world's soil resources, for improvement of their productivity, and for their conservation for future generations. The basic principles of the Charter are still valid today, but need in some parts some revision and updates in light of the new scientific findings and the rapidly changing environmental and social conditions of the world. A new updated version of the Charter will be submitted for endorsement in 2015.
3. Compilation of the new World Soil Resources Report (WSRR). This report is intended as the flagship publication of the ITPS and should be aiming towards the establishment of a fully updated assessment of the status and trends of the global soil resources. It should be forming the reference for future policy decisions at global scale concerning soils and will provide an updated assessment of soil degradation processes after the last Global Assessment of the Status of human-Induced Soil Degradation (GLASOD) of 1990. It is planned to release the new WSRR in 2015, 25 years after GLASOD and in occasion of the UN International Year of Soils 2015.
4. Endorsement of the plan of action for the 5 pillars of the GSP. As one of its main institutional tasks, the ITPS is mandated by the GSP plenary to develop and finalize the detailed plans of action for the five main pillars of the GSP. The plans of action form the basis for the GSP activities and require a very detailed analysis of the current status and the necessary immediate actions on the priority areas identified by the GSP partners. The draft plans of action are prepared by specific working groups established by the GSP Secretariat that submit then the

first draft versions to the ITPS for finalization. Once finalized and endorsed by the ITPS the GSP Secretariat (FAO) submits the five action plans to the GSP Plenary for final formal approval and subsequent implementation. Donors are encouraged to provide the necessary financial support to their implementation. A specific trust fund has been established by the GSP to facilitate the financial support by potential donors.

5. Establishing relations with other science-policy interfaces like IPCC, IPBES and the SPI of UNCCD. The ITPS has the ambition to provide policy relevant scientific advice and information not only top the GSP, but also to other related UN bodies and conventions. It is well recognized by the climate change community that soils play a crucial role in the global greenhouse gas balance, especially as the major terrestrial carbon sink. Soil related scientific knowledge and information is needed by the UNFCCC to properly take decisions in relation to soils. The need for such data and information clearly emerged in the recent debate around the LULUCF issue. The special attention that this convention is now dedicating to organic soils, like peatlands, will require a substantial scientific input in the near future. Similar requirements are emerging also in other policy areas, like biodiversity and desertification. The recently established IPBES will require substantial support on soil related issues, especially on the recently decided fast track assessment on land degradation. The same applies for the recently established science-policy interface (SPI) of the UNCCD. The ITPS is open to collaborations with all these various organizations and stakeholders and formal contacts are on going between the GSP Secretariat and the various bodies on this matter.

3. Regional Soil Partnerships

The actual implementation on the ground of the various plan of action of the GSP will have to be coordinated at Regional level. Soil properties and drivers and pressures are very different in the various parts of the world. Therefore there cannot be "one solution fits it all" approaches. Each action needs to be tailored and adapted to the local conditions. Therefore it is of crucial importance to activate Regional partnerships that can mobilize the existing stakeholders and organizations at national and local level. Seven Regional Soil Partnerships are proposed, following the FAO regional structure. But it is already evident that sub-regional partnerships will have to emerge in order to get closer to the actual realities at local level. Regional and Sub-Regional Partnerships have already been established for Europe, Eurasia, Asia, Near East and North Africa, West and Central Africa, East and South Africa, Mexico, Central America and the Caribbean, South America (Figure 3). More will follow in the near future, as the implementation of the GSP will get more and more operational.

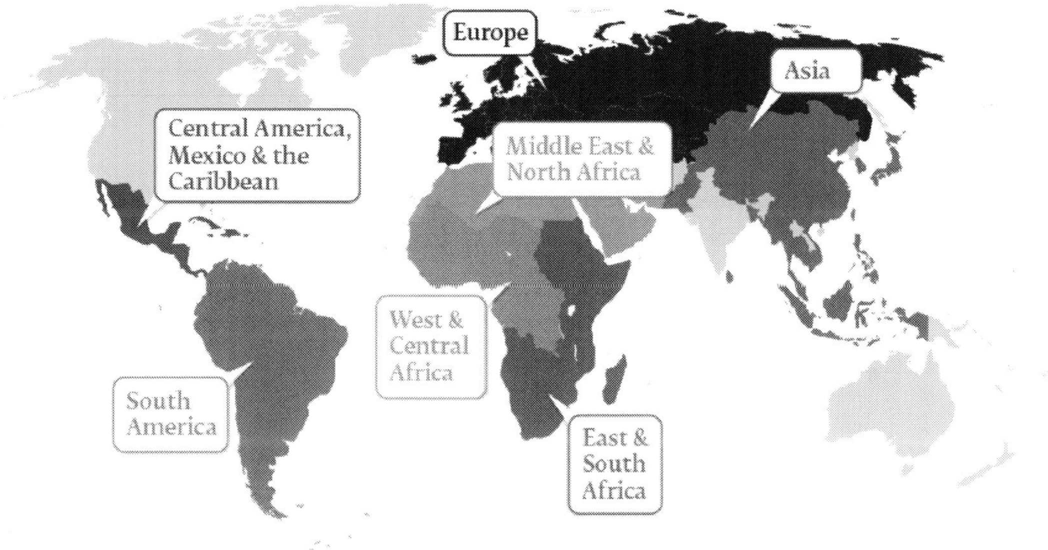

Figure 3. Current areas of world covered by Regional Soil Partnerships (status February 2014).

4. Conclusions and way forward

The Global Soil Partnership is by now in its full operational phase. In less than two years an effective voluntary partnership has emerged facilitating all those stakeholders and governments that are genuinely committed to soil protection and sustainable soil management. This initiative has proved once more that if there is the will, actions can be implemented also at international and global level. Compared to lengthy processes in other more formal frameworks, this simple voluntary partnership was able to get into operational mode in a very short time. All its planned bodies and structures are in place and fully operational, its action plans are in their final development phase and are soon ready for full implementation and substantial financial means have been already mobilized by donors. There is great hope that the GSP will trigger the needed pro-active thinking by all involved organizations towards achieving measurable progress in a short time frame on the vital issue of protecting our limited soil resources for future generations. Failing to do so will have catastrophic consequences for all of us. Recognizing that soil resources are of crucial importance for food security, climate change, biodiversity and many other ecosystem services is a first step towards soil protection and sustainable soil management. The recent endorsement by the UN General Assembly of the UN World Soil Day (5th of December) and the UN International Year of Soils 2015 is a major step towards building the framework for global awareness raising initiatives on this crucial issue. As a first achievement of the GSP, this is already an outstanding result of this very young, but already very effective partnership.

References

[1] Montanarella L, Vargas R 2012 Global governance of soil resources as a necessary condition for sustainable development, *Curr Opin Environ Sus* **4** 559-564

Policy and effective action for soil security: a need for reframing the soil story

Johan Bouma[1]
[1] em. Prof. Wageningen University, the Netherlands.

E-mail: johan.bouma@planet.nl

Abstract. Soil science is a vital scientific discipline producing cutting-edge research in its various sub-disciplines. Its role in studying major environmental issues, often defined as food security, water and energy availability, climate change and biodiversity loss, is, however, less obvious to stakeholders, policy makers and the public at large. It deserves better. Reframing the soil story by taking a pro-active interdisciplinary approach in demonstrating the role of soils when studying these issues is advocated in the context of the ten Sustainable Development Goals. Soil change matters not only in a negative way when referring to erosion and degradation but also, and particularly, to soil improvement. But only successfully completed programs in practice will be convincing and that's why an active role of soil researchers and soil scientists, acting as knowledge brokers, is advocated in transdisciplinary programs. This includes more emphasis on preparation and implementation than is allowed in current programs. A case study is presented and a *narrative* is used to link the five major environmental issues in a logical sequence, showing their interdependence. Current soil research and education programs should reflect demands made by inter- and transdisciplinary approaches and need a new, fresh approach.

1. Introduction

The title of this conference is well chosen. Mankind is facing a number of major environmental challenges in the coming decades as the world population is likely to exceed 9 billion. Strategic reports on the state of the environment usually list food security, freshwater and energy shortages, climate change and biodiversity loss as key environmental issues. The preservation of soils, as such, is usually not listed as a key issue but the way the soil resource is changing as a result of human (mis) management certainly is such a key issue, because it has major effects on each of the five recognized environmental issues mentioned above. Perhaps soil science has been too much inward looking, paying inadequate attention to documenting these major effects. Also, when scanning soil science literature, emphasis tends to be on erosion,- degradation and pollution of soils, representing soil change in the wrong direction. But soil research has also been quite successful in developing soil conservation and soil improvement techniques that can turn around degradation processes by applying innovative soil management. Even though much knowledge and expertise is available, soil degradation in various forms still proceeds at an alarming scale. The question is, 'How to raise more awareness and achieve soil change in a positive direction?' The objective of this paper is to discuss this issue by analysing: (i) the way in which various stakeholders, policy makers and citizens-at-large experience environmental issues and concerns; (ii) the role soil science plays in framing these issues; (iii) ways to effectively participate in inter- and transdisciplinary environmental research programs, and (iv) a case study where an attempt was made to apply some of the ideas being propagated here.

2. Environmental awareness in society

Environmental awareness has increased considerably during the last 50 years, starting with the landmark book by Rachel Carson: "Silent Spring" in the 1960's, the Club of Rome in the 1970's, emphasis on Sustainable Development following the Brundtland report in the 1980's and the activities of the IPCC, studying climate change, also since the 1980's. The, at least partly successful, UN Millennium Goals of the 2000's have now been succeeded by the UN Sustainable Development Goals (SDG's) and Targets for 2030. They reflect a broad societal context in which to consider environmental issues. When discussing soil change, I would propose we do so in the broad context of the SDG's. The 17 Goals (as established July 2014) can be summarized as follows (http://sustainabledevelopment.un.org):

1. End poverty in all its forms everywhere
2. End hunger, achieve food security and improve nutrition and promote sustainable agriculture
3. Ensure healthy lives and promote wellbeing for all ages
4. Ensure inclusive and equitable quality education and promote life-long learning opportunities for all
5. Achieve gender equality and empower women and girls
6. Ensure availability and sustainable management of water and sanitation for all
7. Ensure access to affordable, reliable, sustainable and modern energy for all
8. Promote sustained, inclusive and sustainable economic growth, full and productive employment and decent work for all
9. Build resilient infrastructure and promote inclusive and sustainable industrialization and foster innovation
10. Reduce inequality within and between countries
11. Make cities and human settlements inclusive, safe, resilient and sustainable
12. Ensure sustainable consumption and production patterns
13. Take urgent action to combat climate change and its impacts
14. Conserve and sustainably use the oceans, seas and marine resources for sustainable development
15. Protect, restore and promote sustainable use of terrestrial ecosystems, sustainably managed forests, combat desertification and halt and reverse land degradation and halt biodiversity loss
16. Promote peaceful and inclusive societies for sustainable development, access to justice for all and build effective, accountable and inclusive institutions at all levels
17. Strengthen the means of implementation and revitalize the global partnership for sustainable development

Each goal is specified by three specific targets, that are not mentioned here. It is easy to be cynical about what can be seen as unrealistic, wide ranging "shopping lists". Still, they provide a context and specific targets and this is particularly valuable, it would seem, for a relatively small profession as soil science that appears to suffer from a lack of recognition in terms of their role in studying the major environmental problems of the immediate future. Earlier, in the introduction of this paper, key environmental issues that are being recognized in international strategic documents were mentioned. They all come back here, be it in an attractive broad socio-economic context. Food security is key in goals 2 and 12, while of obvious importance for other goals. Water is mentioned in goal 6, climate in goal 13, biodiversity on goal 15 and energy in goal 7.

In summary, the suggestion is to frame our activities in the context of the SDG's which offer a development context for the world up to 2030. This avoids looking back feeling sorry for ourselves as not being recognized, but putting emphasis on what can and must be done in future. The potential for soil science is immense.

3. Framing soil science in the SDG context

Earlier, Bouma and McBratney [1] and Bouma [2] have emphasized the need to more clearly demonstrate the important role that soils can play to ensure food security and freshwater availability and to combat climate change, energy shortages and biodiversity loss. This is not a difficult challenge: no soils, no crops. Soils are very effective in purifying percolating wastewater thereby protecting aquifers. Next to the oceans, soils contain more Carbon (C) than all the vegetation of the world combined and increasing soil-C is a major route for climate change mitigation. Biofuels grow on soils and improved agricultural practices, avoiding losses of agrochemicals to air and water, are crucial for biodiversity. Rather than laying emphasis on the importance of soils as such, it helps to emphasize the importance of soils for crop production, water purification, climate mitigation and biodiversity preservation. But only talking about this or presenting conceptual and theoretical papers is not going to carry the day. Only specific examples out there in the field that demonstrate successful efforts will be convincing. We need more of those or, better, we should more effectively communicate the many successful narratives that are already there.

The SDG's, as mentioned above, add an additional socio-economic dimension to restricted environmental discussions and it would be wise to keep the SDG's in mind when reporting soil studies. For example, female farmers are dominant in many developing countries and agricultural development will therefore be favourable for gender equality (goal 5). When considering the SDG's, the various roles played by agriculture are quite prominent. Agriculture is crucial to end hunger and extreme poverty, it strongly affects the planetary boundaries of Rockstrom et al [3], population health, climate change, ecosystem services and governance, the latter if only by ownership issues related to land. This discussion is important for soil science as some soil scientists feel that the profession has been too closely related to agriculture in the past and advocate detachment from agriculture and a closer association with the geosciences. This can be done, but when striving for an increased societal profile for the profession by reframing soils, a link with the SDG's would be wise and that implies, as discussed, the need for a remaining focus on agriculture.

Finally, Bouma [2] pointed out that the five key environmental issues, discussed above, should not be seen in isolation: improving food production (the first issue among equals) may have positive effects on the other four issues and he advocates development of narratives in which all the issues are linked in a logical manner.

4. How to reach out to other disciplines and with stakeholders?

4.1. Interdisciplinarity

Too many papers and strategic documents contain abstract pleas for inter- and transdisciplinarity, the first term relating to work with other disciplines and the second to interaction with various stakeholders and policy makers. Of course, the SDG goals cannot be reached with disciplinary research alone but only talking about the issue of interdisciplinarity is not enough and we should face up to what this really means. It helps to realize that the hydrological, climatological, agricultural and ecological research communities are tightly organized, sometimes resembling closed shops, just like soil science. They, like us, have their own journals, meetings and research programs. Funds are scarce everywhere and colleagues from adjacent disciplines are only being invited as members of interdisciplinary research teams if they can make crucial contributions. The challenge, therefore, is to produce such contributions. How to do that? In order to make predictions of future conditions, the use of simulation models is essential in all disciplines dealing with major environmental issues. Such models are gross simplifications of reality and simplifying input of adjacent disciplines is easier accomplished than simplifying your own input. Soil scientists are often not members of the interdisciplinary teams and are therefore easy victims of simplification. The approach to take by soil scientists needs more discussion but should go beyond expressing an abstract and conceptual need for interdisciplinarity. I would suggest (see also [4] and [5]) to run their models with and without good soil data and compare the results. For example, many current hydrological basin-models do not contain soil data or use it only in an elementary form. (e.g. Droogers and Bouma, 2014). The hypothesis would be that using up-to-date soil data produces better results. But the verdict is still out as there appear to be no studies so far taking this approach. And important: use their models and not your own! Models are a symbol of scientific virility and recognizing this will be helpful in communicating with colleagues in other disciplines.

4.2. Transdisciplinarity

Working with stakeholders, citizen groups, NGO's and policy makers is complicated (see also [6]). The old paradigm of scientists studying self-defined problems and producing "solutions" that are passed on to grateful citizens to be applied in practice is not realistic anymore, nor has it ever been. The relation between science and society is changing dramatically in the 21st century if only because of modern information technology. Lessons were learned in a major research program in the Netherlands on Sustainable Agriculture, as reported by Bouma et al [7]. The SDG's and their specific targets present "wicked" problems for which no single, "magic" solutions exist. Complicated and cumbersome compromises have to be found balancing economic, social and economic considerations.

Two major conclusions were reached in the Dutch study:

(i) before starting any project, more time should be spent in defining stakeholders involved and their opinions, interests and hidden motives. Such motives are often only expressed over a beer rather than in formal settings. As is, researchers often spend most of their time on formulating project proposals with minor, superficial contacts with stakeholders, followed by a quick start once the project has been funded and preparations for the next project while the former one is still in progress;

(ii) what really counts in the end are specific results. Just delivering reports and writing papers is not enough. The important role of "knowledge brokers" was identified in trying to realize implementation of plans. They are, ideally, members of the scientific team and have a high social intelligence being able to make the right injections of knowledge at the right time and place and manner to move things along. As is, we often ignore this implementation phase and move on to the next project. As many , if not all, environmental issues are land related, soil scientists would appear to be in an excellent position to act as "knowledge brokers", the more so since soil surveys in the early phase of the profession involved extensive contacts with land users.

5. A case study reflecting an inter- and transdisciplinary struggle

In the early 1990's dairy farmers in the Northern Frisian Woodlands (NFW) in the Netherlands challenged environmental legislation on ammonia emission from manure spread at the soil surface, that was believed to have adverse effects on nature quality in adjacent nature areas. The NFW area was designated as a "national landscape" because of its small scale character with hedgerows separating relatively small, elongated fields. To reduce emissions, farmers were required to inject liquid manure in the soil and they did not like that because they expected damage to soil structure and loss of control because the heavy machinery to be used was operated by independent contractors.

Besides, operating large machines on small fields offered operational problems. They proposed an alternative by feeding their cows a low-protein diet resulting in manure ("new" manure) with less ammonia, as had been proved to be feasible elsewhere [8]. This way they could realize comparable ammonia emissions as found when injecting manure. However, this was against the law and several farmers received fines of thousands of euros when they refused to inject their manure. This resulted in a decade-long struggle with legislators and researchers as reported by Bouma et al [4] as case study no.1.

In the late nineties the government allowed experimentation for a limited number of farmers but results were inconclusive, partly because of disagreements among researchers. As time went by, the farmers increasingly broadened the scope of their activities, embracing a "cradle-to-cradle" approach that was studied in 2010 with Life Cycle Analysis (LCA) [9]. The "cradle-to-cradle" approach included use of less chemical fertilizers and external inputs and an increased application of animal manure, produced on-farm.

A dramatic point was reached in July 2013 when the government decided that experiments would be terminated and that farmers had to abide by the existing law. Following an intensive political campaign, all parties (!) in the Dutch Parliament approved a motion asking the government to extend the experimental period for another five years for a large number of farmers, allowing them to apply their "new" manure at the surface. This may result in a defacto approval of the cradle-to-cradle approach, including surface application of "new" manure, and illustrates both the importance of this issue on national level and the impact of successful lobbying.

The LCA study compared seven cradle-to-cradle farms (C) with seven comparable farms using more traditional procedures (T). In summary, both populations had very large standard deviations but results showed that:

(i) C had 20% higher soil organic matter contents (186 versus 156 tons/ha) partly because meadows were not plowed and reseeded every five years as were most of the T meadows. The difference was statistically significant;

(ii) C used 15% less energy than T (4.9 MJ/kg milk versus 4.3MJ/kg) because they applied their manure themselves rather than engage contractors and they used less chemical fertilizer. The difference was statistically significant;

(iii) Values for global warming-, acidification- and eutrophication potential were lower for C but not significantly so. Also application rates for chemical fertilizers were lower for C (128kg N/ha versus 146 kg N/ha) but this difference was not statistically significant either. Average measured nitrate contents of groundwater were 12 mg/l for (C) and 22 mg/l for (T). No significant difference and in any case both below the EU threshold of 50 mg/l, and

(iv) Average income for C farms was 40% higher than for (T) but standard deviations were very high and the difference was not significant. C farmers received subsidies for maintaining their hedgerows, which contributed towards their higher income.

The NFW study allows a *narrative*, as mentioned above, in terms of an efficient production system of milk providing a higher income to farmers, partly because of lower production costs, while *water quality* is improved, and the soil organic matter content increases, providing a contribution to *climate mitigation*. *Energy use* was significantly lower and lower emission of nitrogen compounds was favourable for *biodiversity* in surrounding nature areas. In addition, hedgerows, that were maintained in the (C) system, contributed significantly to biodiversity.

In summary, all major environmental issues, as mentioned in the introduction, are being served by the (C) system under the overall banner of food production and this conclusion can be extended to the SDG's by recognizing that implicitly contributions are made to stay within the planetary boundaries of Rockstrom et al [3] (Goals 2,6,7,13, 14,15), that rural prosperity has increased (goal 8), while serving goals 7 and 13 in terms of climate and energy and goal 15 in terms of biodiversity, water and soil. Goal 17 is intriguing. The NFW study showed government to be highly risk averse and top-down, unable to respond to bottom-up initiatives that, if embraced in a timely manner, would have supported government objectives. Also, concepts of environmental indicators, thresholds and proxies were poorly defined, requiring more transparency (e.g. [10]).

Indicators for groundwater quality are clear: nitrate concentrations are defined by a threshold of 50 mg/l. However, as little monitoring data were available around 1990, a proxy of 170 kg N/ha was defined. This was fine at the time, but now many monitoring points provide lots of data on groundwater quality, if only because of the availability of modern sensors, and there is good reason to move away from indirect proxies and focus on the threshold itself. The ammonia story is more complicated. The indicator is defacto deposition of N in nature areas, which is hardly measured up to the present time and is calculated with models. The Government has only defined a national emission threshold of 128 kton ammonia, but this not regionally specified. The proxy of manure injection is intended to reduce ammonia emission from the field and measurements show that this can be achieved. But how about deposition, the real threshold? The research community was divided and provided poor advice to regulatory agencies. In retrospect, more transparency and clarity of objectives could have resulted in better governance. Farmers are willing to follow environmental rules because a good environmental quality is a key element of their operations. But when rules are unclear and cannot be explained, their resistance is justified and should be taken seriously.

But overall, "soil change matters" and change was clearly positive here because soil quality has improved as the higher organic matter content leads to a higher moisture supply capacity, better filtering capacity and higher soil biodiversity.

6. Lessons learned

Soil science is a vital scientific discipline producing cutting-edge research in its various sub-disciplines. Its role in studying major environmental issues, often defined as food security, water and energy availability, climate change and biodiversity loss, appears to be, however, less obvious to stakeholders, policy makers and the public at large. This should be improved because the societal relevance of soil science is not as high as it could and should be. The following suggestions are made that can contribute to the necessary reframing (see also [1], [2]):

(i) Taking a pro-active interdisciplinary approach in demonstrating the role of soils when studying the five major environmental issues, as advocated in the context of Sustainable Development Goals.

(ii) Taking an active role in transdisciplinary efforts by developing a role as *knowledge broker*. Spend more time in preparing research programs by seriously involving stakeholders and stay involved in the implementation phase.

(iii) Reconnecting the knowledge chain, linking tacit knowledge and field experience to basic soil research in both directions. This link was well developed in the past when soil surveyors were in close contact with researchers but appears to be broken now. A transparent knowledge chain can also be effective in improving environmental regulations by explaining relations between indicators, thresholds and proxies, as discussed in the case study.

(iv) Judging scientific quality not only by current quality indicators primarily based on publications in international scientific journals but also by its societal relevance, reflecting current discussions about the relation between science and society in the 21st century. (www.scienceintransition.nl)

(v) Reflecting requirements of inter- and transdisciplinarity in soil education and - communication.

(vi) Showing that Soil Change Matters by not only emphasizing negative changes, following erosion and degradation, but also by showing successful applications of innovative management leading to positive changes in soil quality. Emphasize the role of and the impact on real-life people when describing case studies in terms of a narrative. Only real examples in the field, such as the NFW study, are convincing. There are many more to be told. Shout them from the mountain top!

A final comment on the question whether or not it is wise to keep emphasizing the role of soils while people at large, policy makers and entrepreneurs do not recognize our subtle differences between the terms soil, land or, it must be said,: "dirt". The word "land" touches a nerve with many people: "This Is My Land!". We talk about: "land evaluation" when defining land use potentials, not: "soil evaluation". (e.g. [11]). Still, land evaluation is a broad concept involving economic and social aspects as well as the impact of many other scientific disciplines than soil. A continued focus on soil is attractive and needed to establish and maintain our role in the overall studies on land evaluation. But we should always keep the broader picture in the back of our minds when facing our daily grinds.

Acknowledgements

The author acknowledges financial support of SKB (Sustainable Soil Development) for the NFW study.

References

[1] Bouma J and McBratney AB 2013 Framing soils as an actor when dealing with wicked environmental problems. *Geoderma* **200-201** 130-39

[2] Bouma J 2014 Soil science contributions towards Sustainable Development Goals and their implementation: linking soil functions with ecosystem services. *J Plant Nutr & Soil Sci* **177** 111-20
[3] Rockstrom J et al. 2009 A safe operating space for humanity. *Nature* **461** 472-475
[4] Bouma J, Van Altvorst AC, Eweg R, Smeets PJAM, Van Latesteijn HC (2011) The role of knowledge when studying innovation and the associated wicked sustainability problems in agriculture. *Adv Agron* **113** 285-314. Academic Press. USA
[5] Droogers P, Bouma J 2014 Simulation modeling for water governance in basins. *Int J Water Resource Dev* **30(3)** 475-494
[6] Sayer J, Sunderland T, Ghazoul J, Pfund JL, Sheil D, Meijaard E, Venter M, Boedhihartono AG, Day M, Garcia C, Van Ooster C, and Buck LA 2013 Ten principles for a landscape approach to reconciling agriculture, conservation and other competing land uses. *Proc Nat Acad Sci* **110** 8349-56
[7] Bouma J, Droogers P, Sonneveld MPW, Ritsema CJ, Hunink JE, Immerzeel WE, Kauffman S 2011 Hydropedological insights when considering catchment classification. *Hydrol Earth Syst Sc* **15** 1909-19
[8] Sonneveld MPW, Schroder JJ, De Vos JA, Monteny GJ, Musquera J, Hol J, Lantinga MJ, Verhoeven F and Bouma Jand Bouma J 2008 A Whole-Farm Strategy to Reduce Environmental Impacts of Nitrogen. *J of Env Qual* **37** (3) 333-37
[9] Dolman MA., Sonneveld MPW, Mollenhorst H and de Boer IJM 2014 Benchmarking the economic, Environmental and societal performance of Dutch dairy farms aiming at internal cycling of nutrients. *J Cleaner Production* **73** 245-52
[10] Bouma J 2011 Applying indicators, threshold values and proxies in environmental legislation: A case study for Dutch dairy farming. *Env Sci and Policy* **14** 231-38
[11] Bouma J, Stoorvogel JJ, Sonneveld WMP 2012 Land Evaluation for Landscape Units. Handbook of Soil Science, Second Edition. P.M.Huang, Y.Li and M.Summer (Eds). Chapter 34. P.34-1 to 34-22. CRC Press. Boca Raton. London. New York

The policy and science of soil change – a Victorian perspective

Jane Fisher[1] **and Michael C. Crawford**[2]

[1] Agriculture Policy, Department of Economic Development, Jobs, Transport and Resources, PO Box 500 Melbourne, Victoria, 8002, Australia
[2] Agriculture Research, Department of Economic Development, Jobs, Transport and Resources, PO Box 3100 Bendigo DC, Victoria 3554, Australia

E-mail: jane.fisher@ecodev.vic.gov.au

Abstract. Understanding and managing soil change is an important component of maintaining soil health and soil security which is important for the future of agricultural productivity in Victoria. Historically, soil policy in Victoria has been dealt with on the basis of a single issue. With the emergence of farming systems thinking, and the concept of soil health and soil security, a more holistic approach is now being taken. A seven-step policy framework has been developed that promotes dialogue between scientist and policy makers. The questions it asks (what is the problem and how can it be solved?) clarify the role of government investment, and developing partnerships between science and policy, enables early identification of potential policy problems and development of appropriate policy interventions to manage soil change and ultimately soil health, soil security and soil productivity.

1. Introduction

Soil is formed extremely slowly over centuries from the interaction of biotic activities on the original land surface under the influence of climate, topography, and weathering processes. Soil is constantly being modified under the influence of these natural processes and by human use. Soils are complex ecosystems that form part of or are intimately linked with other ecosystems (such as grasslands or forests).

People have long valued soil mostly for its role in facilitating plant (including crop) growth. The Millennium Ecosystem Assessment [1] more broadly classified the value of all ecosystems to humans through the services they provide as provisioning, regulating, supporting and cultural services. In addition to the recognised role of soil in provisioning services (e.g. crops) it has a fundamentally important direct and indirect role in the delivery of the other services, very often achieved concomitantly. For example, soil can simultaneously sequester carbon, store water and nutrients, and support decomposition and recycling processes [2].

Human activities can greatly influence the services soil can provide. While some of these services provide private benefits, others benefit the public at large. In view of this, and considering the overall value of these benefits, soil management has been an important aspect of public policy, globally, nationally and regionally (including the Australian state of Victoria) for many decades, to both maintain the public benefits and to assist in enhancing private benefits (i.e. via agriculture and forestry).

The term "soil health" is widely used in terms of productive agriculture [2]. Bennett et al. [3] re-examined the concept of soil health in the context of Australian agriculture and found knowledge gaps in areas related to public, rather than to private, good. More recently, McBratney et al. [4] proposed

IOP Conf. Series: Earth and Environmental Science **25** (2015) 012003 doi:10.1088/1755-1315/25/1/012003

soil security as a more encompassing concept that should include soil health as one of its parameters. They defined soil security as being "concerned with the maintenance and improvement of the global soil resource to produce food, fibre and fresh water, contribute to energy and climate sustainability, and maintain the biodiversity and overall protection of the ecosystem". Maintaining and improving agricultural productivity depends on good soil health and relies on maintaining soil security. From a state government perspective, declining soil security in Victoria would adversely affect agricultural productivity and other ecosystem services, with a negative financial impact on both Victorian and global food security.

The Government of Victoria recognises that Victoria's soils are vital to the continuing environmental, social and economic prosperity of the State [5, 6]. The government also recognises that healthy soil is the foundation of many key ecosystem services, including the protection of land, water and biodiversity, and contributes directly to the habitat for Victoria's celebrated flora and fauna [6]. Victoria's soil currently generates over $9 billion worth of agricultural exports each year, accounting for 29% of Australia's total agricultural exports from just 3% of Australia's land mass, and enhancing regional and local economies across the state [7].

The Victorian Government has legislative requirements for reporting on change in soil condition under the *Catchment and Land Protection Act (1994)* by the Victorian Catchment Management Council and the *Commissioner for Environmental Sustainability Act (2003)* by the Commissioner for the Environment. Both report every five years. This reporting is important in providing measurements of change in soil condition that might trigger government intervention. However, although there is a requirement for regular reporting there is no systematic funding program for the monitoring and measurement of soil condition [8]. Hence, the methodologies used and sites sampled are not necessarily consistent between reporting periods making valid comparison difficult.

Victoria's soil resources provide the basis for generating many benefits and make a significant contribution to Victoria's economic, environmental and social wellbeing. These benefits are both private and public in nature, and are trans-generational, in that the benefits exist for both current and future generations. As such, there are many parties that have a stake in the sustainable use, management and protection of Victoria's soil resources and the interests of these parties can sometimes be in conflict. These situations justify government considering intervention and when there is government intervention, or the possibility of it, there arises a need for policy to guide that intervention.

Roughly 62 percent of Victoria's land is managed privately, and nearly 90 percent of this is in some form of agricultural land use. As well as its lead role in the management of public land, the former Department of Environment and Primary Industries (DEPI)[3] had a lead government agency role for management of soil health on private land. By comparison with other Australian jurisdictions, DEPI made significant investments in soil health in policy, research and practice change [9].There are strong linkages between policy, research and practice change in that all three components inform each other. The focus of this paper is the way in which research (or science) informs policy, and vice versa, for agricultural soils in the Victorian context.

2. A brief history of Victoria's soil policy and governance arrangements

Victorian government policy, requiring implementation through investment and action, in relation to soil change and soil health outcomes has evolved over the decades in response to different drivers and different government philosophies. DEPI was established in 2013 through the merger of the former

[3]Following the November 2014 Victorian State election, DEPI was divided between the Department of Economic Development, Jobs, Transport and Resources, and the Department of Environment, Land, Water and Planning. Agriculture has moved to The Department of Economic Development is responsible for agriculture, which includes soil on private land.

Department of Primary Industries (DPI) and Department of Sustainability and Environment (DSE). Both of these Departments had a role in soil health outcomes, which in turn partly reflected the predecessor organisations from which they arose (VAGO, 2010). From 1st July 2013, government responsibilities for soil health was shared between DEPI, the 10 Catchment Management Authorities, the Environment Protection Authority Victoria (EPA Victoria) and Parks Victoria. DEPI is the principal agency responsible for developing soil health policy. Other agencies, including the Victorian Catchment Management Council (VCMC) and Commissioner for Environmental Sustainability are responsible for reporting on the management and condition of land and water assets including soil. The VCMC also has a legislative role in advising the Minister on catchment matters including soil.

The former Department of Primary Industries (and historically the Department of Agriculture) always had a focus on supporting productive agriculture, and developing soil management practices that supported productive crop and pasture production. In doing so, there was often a focus on addressing production constraints, and various issues arose and fell in prominence over time (i.e. trace element deficiencies (1950s) [10], acidification [11] and waterlogging [12] salinity (1980s) [13]), soil structure (1990s) [14] subsoil constraints (2000s) [15,16,17], soil organic matter (2010s) [18,19,20].

The former Department of Sustainability and Environment (and historically the former Soil Conservation Authority) had a greater focus on soil protection, land degradation and water quality issues, partly from a productivity (private good) perspective, but just as importantly, from an environmental (public good) perspective. The soil erosion focus of the 1950s, a consequence of wind and water events which led to the formation of the Soil Conservation Authority [21], was superseded by the salinity focus of the 1990s [22] and the concurrent Decade of Landcare [23, 24] and subsequent Salinity Action Plans [25, 26]. The Government of Victoria created Catchment Management Authorities (CMAs) in the mid-1990s to be important delivery partners and instruments of government intervention at a local and regional scale. The CMAs operate in a regional partnership structure with Department of Environment, Land, Water and Planning, the Environment Protection Authority, Parks Victoria and local councils as well as local stakeholders and land managers. State and federal funding for CMAs is directed to on-ground soil health works on private and public land and stream and rivers, and to whole-farm planning activities on private land. Funds are invested through CMAs according to the priorities defined in their respective Regional Catchment Strategies (RCSs).

In the last decade, there has been a policy shift away from addressing individual soil problems to managing soils as complex systems that produce services that benefit society, i.e. an ecosystem services approach [27].

The 2004 Parliamentary Inquiry by the Victorian Parliament's Environment and Natural Resources Committee into the impacts and trends in soil acidity [28] resulted in a step change in government soil policy. This inquiry concluded that acid soils were a major economic and environmental challenge, and that a comprehensive acid soils management strategy was needed (with associated investment). The government response concluded that it was unwise to focus on a single issue (acidity) as that approach led to a single strategy response (liming), while other issues remained unresolved or unmanaged [29]. The response, informed by close collaboration between former DPI's policy and science staff, established that a better approach was to consider the soil resource holistically by focusing on the issue of 'soil health'. The response postulated that the underlying causes of decline in soil health on private land, their impact on productivity, and interactions between soil, water and wind are often extremely complex and variable. It also recognised that there is no universal formula for managing soil health and that an adaptive management approach was needed [30].

These conclusions led to the development of former DPI's Soil Health Policy Framework for Productive Agriculture [31]. The purpose of this framework was to clarify the role of DPI in soil management for productivity on private land and to encourage the practical application of sound policy principles to making investment decisions. Along with the former DSE Soil Health Strategy [32], it continued to set the basis for DEPI's interventions and investment in soil and soil health related research, extension, and policy. The regulatory framework for soil continues to be the *Catchment and Land Protection Act (CaLP) (1994)*, the *Water Act (1989)*, and the *Commonwealth Water Act (2007)*.

Within Victoria, there are other contributors to this area beyond DEPI. A number of CMAs have developed Regional Soil Health Strategies. The last two iterations of Regional Catchment Strategies have tried to establish regional soil priorities (but to varying levels of success). The CaLP Act outlines CMA and VCMC reporting responsibilities that could have a key role in tracking soil change and evaluating the success of policies. Similarly, the Commissioner for Environmental Sustainability has a role in SoE reporting. The CaLP Act also outlines the formal role of the VCMC in providing independent advice to the Minister on land and water issues and makes provision for the potential role of "land management notices" requiring a landholder to undertake actions that prevent unreasonable amounts of soil loss from impacting on neighbours or adjacent assets). All of these influence, or provide the context for, soil policy and governance arrangements on agricultural land.

3. The current context

The historical analysis shows that government intervention in soil management has been strongly influenced by the need to address the issue of the day, or to put it more strongly, crises (e.g. erosion, salinity).

In some cases, government intervention, along with private action, has helped to ensure that an issue once seen as a major problem is less significant now than it was in the past. A key example of this is erosion. Research by DPI that measured the adoption of conservation farming practices in northwest Victoria reveals that the proportion of paddocks under conservation management nearly doubled from around 44 per cent in 1996 to over 82 per cent in 2009 (and up from 0 per cent in the 1950s) [33]. This has greatly reduced the risk of wind erosion from these lands, and the occurrence of dust storms is now much less than it was in past decades [33].

In other cases, the crisis was either overstated (maybe because scientific understanding of the problem was limited) or the ability to address the crisis was overestimated. A key example of this is dryland salinity. Two decades ago, the 'rising flood' of dryland salinity was forecast to overtake large areas of productive agricultural land, and widespread efforts were instigated to prevent this (e.g.[34]).Today, in response to improved knowledge of groundwater systems and processes including their response to wetter and drier phases of climate, the forecasts have been revised down significantly, and there is a better understanding of how to manage and influence these processes through soil and land management. The current accepted view is that salinity is a part of the landscape and that Victorian private and public land managers need to learn and have learnt how to manage and adapt to living with it [35]. Government activity, informed by policy and science, now concentrates on realising opportunity and potential for productivity gains. In response to a range of drivers, from November 2010, Victorian government agricultural investment was driven by the policy of doubling food and fibre production by 2030 [36]. Whilst this appeared an audacious goal, analysis showed that it was possible in some sectors, and that the benefits to the economy would be significant. A productive soil resource is recognised as one of the contributing components to achieve this goal.

There is also a realisation by most stakeholders that there are many win-win solutions in soil management. An output from better soil management (e.g. reduced acidity or increased soil organic matter) may lead to improved productivity outcomes, and often also positive environmental outcomes (e.g. increased soil organic matter means carbon has been sequestered in the soil). Further, because the majority of Victoria's soil is under some form of agricultural management, achieving significant environmental outcomes requires interventions to be undertaken in an agricultural context. Government interventions on private land that rely on voluntary action must target productivity drivers to achieve both productivity and environmental outcomes. Where this is not possible, the delivery of environmental outcomes may be encouraged by a system that pays farmers for non-agriculture services.

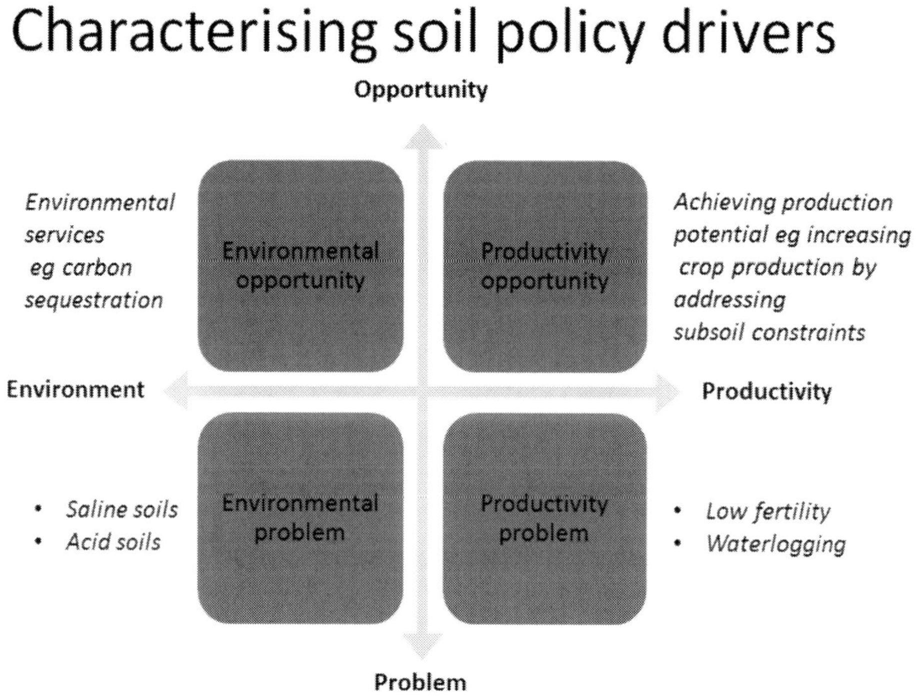

Figure 1: A conceptual model of the drivers of soil policy, showing the ecosystem services continuum from environment through to productivity.

What this history shows is that the drivers for government policy and investment in soil health have changed over time. The major driver at any one time shifts between problem and opportunity and between environment and productivity (Figure 1). Hence, the focus has moved over time from addressing productivity and environmental 'problems' to trying to create 'opportunities', environmental and now productivity, through enhanced soil management. In the agricultural context, maintaining the environment (as defined by soil security), is essential to maintaining and increasing productivity. Cycles change. In the Victorian context, community views are important and influence policy. It is possible that in future, environment will be more prominent than productivity in response to community expectations and environmental reporting will increase, hence increasing the focus and emphasis on monitoring and managing soil change.

4. The future

A key element of future activity in soil science in Victoria is the National Soil Research, Development and Extension (RD&E) Strategy [37] that was launched in March 2014. Developed under the auspices of the National Primary Industries Research, Development and Extension Framework, this strategy brings together the various investor and provider partners in Australian soils RD&E. It has sought to identify what will be some of the 'big' policy issues in Australia and how they can be informed by improved soil data, information and knowledge. In turn, the strategy helps to identify the research priorities, the capabilities needed to address these priorities, and the current state of the nations' capabilities to respond to these needs. From a Victorian perspective, this is a key document in helping to shape the ongoing relationship between soil science and policy. Strategic goals of relevance to Victoria is improving the effectiveness of co-investment to generate and apply new knowledge, improving quality, availability and access to soil data and information and improving communication and exchange of soil knowledge. Other drivers of Victorian agricultural and food policy are the rise in Asia's middle class, coupled with increased access to new markets made available through free trade agreements with Japan, Korea and China, presenting enormous growth opportunities for Australian food and agriculture industries.

The interaction between the National Soil RD&E Strategy and the Victorian government's objectives for food production is under review. One issue to resolve that is relevant to this paper is the high level priority in the National Soil RD&E Strategy on "how to improve the mapping, modelling, monitoring and forecasting of soil-related issues" and how much emphasis is placed on this in Victoria relative to other objectives.

5. What does policy want from (soil) science?

There are many theoretical approaches to the development of public policy, but they largely represent variations on the following seven-step framework developed for the former DPI Soil Health Policy for Productive Agriculture [38] (DPI, 2012):

1. Establish the nature and significance of the problem or opportunity
2. Identify potential market failures
3. Identify objectives and policy options to achieve them
4. Identify implications for private action
5. Assess policy options and developing key performance indicators
6. Identify and design delivery methods and prioritise resources
7. Design and conduct annual reporting and tailored evaluations of the research, development and extension being undertaken to address the problem.

Government can use a variety of instruments and approaches to manage soil health. These include information generation and provision (R&D and extension), regulation, direct government provision of on-ground works, grants and cost-sharing arrangements (subsidies) and other market-based approaches (e.g. fees, levies, trading). Often a mix of these instruments is most appropriate. The distinction between legacy issues and current practices can also be important in determining which mix of policy tools is most appropriate.

Different farming systems, climate, landscape and soil provide the context for prioritising the parameters for measuring change in soil regionally. In Victoria, this is undertaken through the Regional Catchment Strategy process managed by the ten CMAs. To be fully effective, it is necessary to have data, information and knowledge at the appropriate scale. It is arguable that this is not currently the case as the current suite of RCSs all outline that their ability to set spatially relevant priorities around soil change is severely handicapped by a lack of data, information and knowledge at a useful scale and context to their region. Work is being undertaken to improve available data (Mark Imhof, pers comm).

When implemented fully and systematically, this approach represents an evidence-based, defendable approach to government intervention (or absence of) and leads to the effective and efficient use of taxpayer funds. However, there are some caveats to this statement, as often the full achievement of some of these steps, even the first one, is limited by the under, or poor, application of scientific knowledge.

Systematically working through the seven-step framework readily identifies the steps where (in a soils context) scientific knowledge can make a contribution to the policy problem. This process can generate questions such as listed below. Policy wants to have answers to the questions; the questions can be asked at any time, and generally answers are expected as soon as possible. Potential questions include:

- Where are the problems (or opportunities)?
- Are the problems new and emerging, or have they been around for a long time?
- How big (how widespread) is the problem?

- Which industries are affected? What does the soil health problem cost agricultural productivity?
- What is the problem's past, current and future impact on public assets?
- Does a solution exist, or does one need to be developed?
- Is the solution cost effective?
- Is the solution adoptable? What are the barriers to adoption?
- Can data on soil condition be accessed quickly and easily? Can it be interpreted quickly and easily?
- Is a visual (spatial) representation of problems by region and industry readily available?
- How is the problem changing over time? How has it responded to past interventions?

The practicing soil scientist (and science manager) will quickly identify that the answers to some of these questions are not always readily available, often cost money, require time to answer comprehensively, and frequently lead to more questions being asked. The challenge therefore is to identify which of these questions is worth answering; do they require that an answer be developed (i.e. through research and investigation), or are they of a nature that could be readily answered if the data and information was already available (i.e. through monitoring and reporting). An ongoing dialogue between policy advisors and scientists is critical to informing these decisions.

Policymakers also look for consistency of knowledge and the ability of scientists to simplify the complex. They also require clear and precise 'plain english' answers to questions related to quantifying the size of the problem, the cost of solving the problem, and the impact that this will have on either ecosystem services or achieving production potential. Soil scientists, like many scientists, "should strive to: understand the policy maker's perspective, practice excellent communication, be solution oriented, find a champion, avoid appearances of vested interest, and be simple, patient, persistent, resilient, responsive and timely" [39].

6. What does (soil) science want from policy?

Whilst science helps to inform policy, policy helps to inform scientific direction and intent, by providing the strategic context for the research. Scientific funding and effort is driven by clearly identified policy priorities that lead to the achievement of agreed outcomes.

In addition to the strategic context that policy provides for science investment, policy also directly engages with science and asks specific questions of science. What is helpful to science is if policy questions are specific, relevant and answerable (albeit with more research), as opposed to vague, unfocussed and unanswerable. Whilst some questions can be answered immediately, others take time. The identification of 'emerging' policy questions and the issues likely to be of increased concern in two to three years can help science to prepare to answer the specific questions when they need to be answered. Again, an ongoing dialogue between policy advisors and scientists is critical.

7. Conclusions

This paper has attempted to describe the journey taken by soil policy in Victoria over the last sixty years. Soil policy has moved from reacting to the single most important issue of the day to a more holistic, systems based approach. This is achieved by using a 'soil health' and in the future, potentially, a 'soil security' lens to respond to the policy objectives of the government of the day. While the elected government sets the policy agenda, ongoing interaction between soil scientists and policy analysts is required to ensure that the policy can be delivered. Dialogue is also important to enable emerging problems to be identified and managed before they become crises. The policy – soil science conversations need to enable the exploration of the new science, to understand the potential for science to deliver both public and private benefits in the form of ecosystem services, (including productivity) to the State of Victoria. As Bouma [40] observed, soil scientists need to consider changing the focus of their thinking from dealing with threats to concentrating on how they can contribute to achieving the goals of the government of the day. In particular, for the context of this paper and the workshop, a better understanding of critical changes in soil that can affect the economic goals of government, through monitoring, measurement and the application of models in the soil-landscape-farming system nexus.

A well designed and implemented policy framework is an important tool for both parties to use to develop a shared understanding of emerging opportunities and concerns. Clear and concise questions and answers are essential for both parties to work well together.

References

[1] Millennium Ecosystem Assessment 2005 Ecosystems and Human Well-being: Synthesis. Island Press, Washington, DC.
[2] Kibblewhite MG, Ritz K and Swift MJ 2008 Soil health in agricultural systems *Philos Tr R Soc B* **363** no 1492 685-701
[3] Bennett LT, Mele PM, Annett S and Kasel S 2010 Examining links between soil management, soil health, and public benefits in agricultural landscapes: An Australian perspective. *Agric Ecosys Environ* **139**:1-12.
[4] McBratney A, Field DJ and Koch A 2014 The dimensions of soil security. *Geoderma* **213** 203-213
[5] DPI 2012 Soil Health Policy for Productive Agriculture. Department of Primary Industries, State of Victoria, Melbourne.
[6] DSE 2012 Soil Health Strategy – Protecting soil health for environmental values on public and private land. Department of Sustainability and Environment, State of Victoria, Melbourne.
[7] DEPI 2014 Food to Asia Action Plan – Putting Victorian Food and Beverages on Asian Tables Victorian Government, Department of Environment and Primary Industries, 8 Nicholson Street, East Melbourne, March 2014.
[8] Commissioner for Environmental Sustainability Victoria 2013 Victoria: State of the Environment. Victorian Government, Office of the Commissioner for Environmental Sustainability.
[9] Soils Research Development and Extension Working Group 2011 A stocktake of Australia's current investment in soils research, development and extension: a snapshot for 2010-11 Commonwealth of Australia 2011, Department of Agriculture, Fisheries and Forestry, Canberra, ACT
[10] Hosking WJ, I.W Caple IW, Halpin CG, Brown AJ, Paynter DI, Conley DN and North-Coombes PL 1986 "Trace elements for pasture and animals in Victoria", Department of Agriculture and Rural Affairs.
[11] Slattery B and Hollier C 2002 Impacts of Acid Soils in Victoria Department of Natural Resources and Environment.
[12] MacEwan RJ, Gardner WK, Ellington A, Hopkins DG, Bakker AC 1992 Tile and mole drainage for control of waterlogging in duplex soils of south-eastern Australia. *Aust J Exp Agric* **32** , 865–78.
[13] Walker G, Gilfedder M and Williams J 1999 Effectiveness of Current Farming Systems in the Control of Dryland Salinity. CSIRO.
[14] Cockroft B and Olsson KA 2000 Degradation of soil structure due to coalescence of aggregates in no-till, no-traffic beds in irrigated crops. *Aust J Soil Res* **38**(1) 61-70
[15] Armstrong RD, Flood R and Eagle C 2001 What is limiting the water use efficiency of cereals in the southern Wimmera? *Proceedings of the 10th Australian Agronomy Conference*, Hobart, 2001
[16] Rengasamy P (2002) Transient salinity and subsoil constraints to dryland farming in Australian sodic soils. *Aust J Exp Agric* **42** 351-361.
[17] Sadras VO, O'Leary, GJ and Roget, DK 2005 Crop responses to compacted soil: capture and efficiency in the use of water and radiation. *Field Crops Res* **91** 131-48.

[18] Baldock J, Wheeler I, McKenzie N and McBratney A 2012 Soils and climate change: potential impacts on carbon stocks and greenhouse gas emissions, and future research for Australian agriculture. *Crop Pasture Sci* **63**(3):269-283.
[19] Robertson F and Nash D 2013 Limited potential for soil carbon accumulation using current cropping practices in Victoria, Australia. *Agric Ecosys Environ* **165** 130-40
[20] Viscarra Rossel R, Webster R, Bui E, Baldock J 2014 Baseline map of organic carbon in Australian soil to support national carbon accounting and monitoring under climate change. *Global Change Biol* **20**(9) 2953–70.
[21] Thompson GT 1979 A brief history of soil conservation in Victoria. Soil Conservation Authority, Victoria.
[22] The Department of Natural Resources and Environment 2000 Victoria's Salinity Management Framework.
[23] Hawke RJ 1989 Our Country Our Future: Statement on the Environment, Australian Government Publishing Service, Canberra.
[24] National Land and Water Resources Audit, Australian Agriculture Assessment 2001 (a Report of the National Land and Water Resources Audit)(Canberra: National Land and Water Resources Audit, Land and Water Australia on behalf of the Commonwealth of Australia, 2001) http://lwa.gov.au/products/pr010237 and http://lwa.gov.au/products/pr010238 (Accessed 2 September 2014)
[25] Nicholson C, Dahlhaus P, Anderson G, Kelliher C and Stephens M 2006 Corangamite Salinity Action Plan (2005-2008) Corangamite Catchment Management Authority.
[26] Anonymous 2000 Our Vital Resources: A National Action Plan for Salinity & Water Quality, Agriculture, Fisheries and Forestry Australia and Environment Australia: Canberra.
[27] DSE 2012 Soil Health Strategy – Protecting soil health for environmental values on public and private land. Department of Sustainability and Environment, State of Victoria, Melbourne.
[28] ENRC 2004 Inquiry on the Impact and Trends in Soil Acidity, Environment and Natural Resources Committee. Parliamentary Paper No. 59 of session 2003-2004, March. Victoria Parliament http://www.parliament.vic.gov.au/enrc/inquiries/soilacidity/ accessed 1/09/2014.
[29] State of Victoria 2004 the State Government response to the Environment and Natural Resources Committee Inquiry on the Impact and Trends in Soil Acidity http://www.parliament.
[30] Crawford MC, Allaway M, Ryan M 2006 Development of a Soil Health Policy Framework for Victoria, Australia. In '18th World Congress of Soil Science, Philadelphia, Penn. USA'. www.ldd.go.th/18wcss/techprogram/P15678.HTM.
[31] DPI 2012 Soil Health Policy for Productive Agriculture. Department of Primary Industries, State of Victoria, Melbourne.
[32] DSE 2012 Soil Health Strategy – Protecting soil health for environmental values on public and private land. Department of Sustainability and Environment, State of Victoria, Melbourne.
[33] Commissioner for Environment 2013 State of Environment 2013. Department of Environment and Primary Industries.
[34] Victorian Government Salt Task Force 1988 Salt Action: Joint Action - Victoria's strategy for managing land and water salinity.
[35] DSE 2013 Victorian Dryland Salinity Update. Department of Sustainability and Environment, State of Victoria, Melbourne.
[36] DEPI 2014 Food to Asia Action Plan – Putting Victorian Food and Beverages on Asian Tables Victorian Government, Department of Environment and Primary Industries, 8 Nicholson Street, East Melbourne, March 2014.
[37] Commonwealth of Australia 2014 Securing Australia's Soil for profitable industries and healthy landscapes - The National Soil Research, Development and Extension Strategy. vic.gov.au/enrc/inquiries/soilacidity/ accessed 1/09/2014.

[38] DPI 2012 Soil Health Policy for Productive Agriculture. Department of Primary Industries, State of Victoria, Melbourne.
[39] Pannell DJ and Roberts AM 2009 Conducting and delivering integrated research to influence land-use policy: salinity policy in Australia. *Environ Sci Policy* **12**(8) 1088-99
[40] Bouma J 2009 Soils are back on the global agenda: Now what? *Geoderma* **150** 224-5.

IOP Conf. Series: Earth and Environmental Science **25** (2015) 012004 doi:10.1088/1755-1315/25/1/012004

The cost effectiveness of a policy to store carbon in Australian agricultural soils to abate greenhouse gas emissions

Robert E White and Brian Davidson
Faculty of Veterinary and Agricultural Sciences, University of Melbourne, Parkville, Victoria 3010, Australia.

E-mail: robertew@unimelb.edu.au

Abstract. Data for cropping and pastoral enterprises in south eastern Australia were used in a cost-effectiveness analysis to assess the feasibility of abating greenhouse gas (GHG) emissions through storing soil carbon (C) as soil organic matter under the Australian government's Carbon Farming Initiative. We used the C credit value for 2013-14 of $24.15 per tonne of CO_2-equivalent (CO_2-e) and a C storage rate of 0.5 tonne C/hectare/year for conversion of cropland to pasture. Given that a change of enterprise is driven primarily by farmer returns, we found that none of the changes were feasible at current prices, with the exception of wheat to cattle or sheep in an irrigated system, and dryland cotton to cattle or sheep. Given that our model scenario assumed the most favourable economic factors, it is unlikely that increased soil C storage through a change from cropping to pasture can make a significant contribution to abating Australia's CO_2 emissions. However, of greater concern to society is the methane emissions from grazing cattle or sheep, which would negate any gain in soil C under pasture, except for a switch from dryland cropping to sheep.

1. Introduction

The Australian government has a stated objective of decreasing Australia's greenhouse gas (GHG) emissions by 5 percent by the year 2020. Through a Direct Action Policy, the government's objective will be achieved by "*abatement .... purchased via a market mechanism to achieve the lowest price....The lowest cost abatement may be a mix of energy efficiency, cleaning up waste coal mine gas, cleaning up power stations and landfill gas. It may be reafforestation of marginal lands or revegetation or improvement of soil carbon*" [1]. This policy is currently being implemented through an Emissions Reduction Fund (ERF), under the Carbon Farming Initiative (CFI) for which there is a Carbon Credits (Sequestration of Carbon in Soil Using Modelled Abatement Estimates) Methodology Determination 2014 [2]. Currently, this 'determination' sets out possible acceptable 'project types' as being 'sustainable intensification', 'conversion to pasture' and 'stubble retention'.

The modelled abatement estimates for carbon (C) storage are derived from the Australian Government's National Inventory System soil carbon model (the Full Carbon Accounting Model – FullCAM) and the parameters used to generate National Inventory reports for agricultural systems [2]. The estimates currently range from 0.122 to 0.5 tonnes (t) C per hectare (ha) per year (yr), that is, 0.449 to 1.835 t carbon dioxide equivalent (CO_2-e)/ha/yr, the highest figure being for conversion from cropland to pasture. These C storage estimates are consistent with the range of 0.3-0.6 t C/ha/yr reported in a review of national and international literature by Sanderman et al. [3], but higher than reported for intensively managed pastures under dryland or irrigation in recent New Zealand research [4, 5]. However, the estimates are much lower than that claimed for 'biological farming systems' of a

0.15% increase in soil organic carbon (SOC) annually (amounting to 3 t C/ha/yr for a soil of 1% C) [6].

Given the wide disparity between these estimates of SOC storage potential and the fact that farmers will be reluctant to change from a profitable farming enterprise to a less profitable one that may require new skills, capital input, and incur a compliance cost for an approved project, the aim of this paper is to analyze the economic consequences for farmers who change their current farming system to other systems for the purpose of storing SOC. A range of common farming systems and their C storage potentials is considered, as are the changes to farmers' gross margins, the cost of compliance with government policy, and a notional value of a C credit (which as from 1 July 2015 will be set by a 'reverse auction' process). Two measures are used to assess the viability of the program - the cost effectiveness per ha of making a change (a measure of the incentive farmers need to partake), and the cost effectiveness per t C (a measure of what the government might expect from the program). These measures are calculated for the short term (1 yr), reflecting the immediate incentive to change, and the longer term of 25 yrs, as will be permitted under the CFI policy from 1 July 2015. However, it should be noted that projects approved after that date will be subject to a 20% reduction in any credits earned if they are based on a 25-yr 'permanence period' [7].

2. Methods

Cost-effectiveness analysis is a technique that assists in ensuring the efficient use of investment resources in sectors where benefits may be difficult to evaluate [8]. The technique can be used to select amongst alternative projects or practices which have the same objectives (quantified in physical terms). The approach can be used to identify the alternative that minimizes the actual value of costs to provide a given level of output; or conversely, the value of costs that maximizes the output level. The result of the analysis is a cost-effectiveness ratio of outputs to costs that can be used to measure the impact of a policy.

This study is confined to the cost-effectiveness of changing from a cropping to pasture-based livestock enterprise, assuming a C accumulation under grazed pasture of 0.5 t C/ha/yr (amounting to 1.835 t CO_2-e/ha/yr) [2]. Although SOC will approach a new equilibrium value asymptotically, this may take 80-120 years [9] so the assumption of a linear change over the first 25 yr is justified. Although a wide range of cropping systems is possible, for simplicity we have chosen representative irrigated cropping systems in southern New South Wales and dryland cropping systems in northern New South Wales, for which data are readily available [10], and assessed the cost-effectiveness of a change to a cattle or sheep enterprise. The costs are those associated with making a change from one enterprise to another and those of complying with the policy. It is assumed that farmers who undertake the change do so just once and do not opt in and out of the program. This assumption is in line with the analysis undertaken by Cowan et al. [11] who argued that farmers, while being tactically flexible within a production system, persist with their existing production systems despite adverse commodity prices. The 'effect' is the amount of C stored per ha, which measured over time is the difference between the increased SOC under the new enterprise compared with the SOC under 'business as usual'. This effect will often be larger than the difference calculated between SOC under the new enterprise and SOC at time zero (the baseline).

The costs of changing enterprises are simplified by assuming they are confined to a change in the gross margin a farmer receives from his enterprise. This assumes that the capital costs associated with the change are non-existent. In table 1, data on gross margins for various dryland and irrigation cropping and livestock enterprises are presented [10]. The costs of compliance will depend on the costs of auditing the project and reporting, and the costs of SOC measurement to verify FullCAM's performance. At present, these costs cannot be quantified and so the compliance cost has been set at zero in the analysis.

Table 1. Gross margins for irrigated and dryland wheat, soybeans, lucerne, cotton, sheep and cattle enterprises in southern and northern NSW

Enterprise	Soybeans (1)	Maize (2)	Wheat (3)	Lucerne (4)	Cotton (5)	Sheep (6)	Cattle (7)
Irrigated systems							
Farm price ($/t)	625	290	200	310	2400	-	-
Yield (t/ha)	4	11	3.5	15	2.025	10	2.6
Gross revenue ($/ha)	2500	3190	700	4650	5348	1116	1497
Gross Costs ($/ha)	622	1396	442	2249	3167	609	594
Gross Margin ($/ha)	1878	1794	258	2401	2181	903	903
Dryland systems							
Farm price ($/t)	475	240	265	288	1992	-	-
Yield (t/ha)	1.5	3.75	3.5	4	0.6075	2	0.45
Gross revenue ($/ha)	713	900	928	1152	1210	223	183
Gross Costs ($/ha)	393	583	430	781	1160	122	55
Gross Margin ($/ha)	320	317	498	371	50	101	127

Notes: 1. Irrigated - Soybeans (edible) – Beds Murrumbidgee Valley Summer 2012-13. Dryland – Soybeans (no-till) North East NSW Summer 2012-13.
2. Irrigated Maize-Grits (beds) – Murrumbidgee Valley Summer 2012-13. Dryland – Maize (no-till) North East NSW Summer 2012-13.
3. Irrigated Wheat-Biscuit (Flood irrigated contour bay/sod sown) – Murrumbidgee Valley Winter 2012. Dryland – Wheat (no-till, long fallow, after sorghum) North East NSW Winter 2012.
4. Irrigated Lucerne (Flood irrigated, maintenance) – Murrumbidgee Valley Summer 2012-13. Dryland – Lucerne Hay Northern Zone NSW Summer 2010-11.
5. Irrigated Cotton (Roundup ready) – Murrumbidgee Valley Summer 2012-13. Dryland – Cotton (Roundup ready) North West NSW Summer 2011-12.
6. Sheep – Merino Ewes (20 micron) wether lambs finished, December 2012, on irrigated pasture and on dryland (native pasture).
7. Irrigated Cattle – Yearlings Southern/Central NSW (pasture improved) December 2012. Dryland Cattle – Inland Weaners (native pasture).
The sheep enterprise is assumed to be a 1000 ewe flock run at 10 dry stock equivalents (dse)/ha (irrigated pasture) and 2 dse on dryland (native pasture). The cattle herd is assumed to be a 100 head mob, run at 2.6 livestock units (lsu)/ha (irrigated) and 0.5 lsu (dryland).
Source: NSW Department of Primary Industries (2012) Farm Budgets and Costs [10].

These costs and their effects need to be measured against the benefits to a farmer from undertaking the change. The value of a C credit was taken as the Australian value of $24.15 per t CO_2-e for 2013-14, which was used in two ways. First, it is incorporated into the farm-level analysis as a benefit to farmers from undertaking the change, and so is subtracted from the costs of change and compliance. Although it is assumed the C credit is paid for every year of the analysis, it must be prorated up as the difference in SOC accumulated increases over time, in this case over 25 yrs. Second, the C credit value can be used to assess the whole policy, when it becomes the basic price the government has to pay to implement soil C sequestration as part of the policy. In this way, the value of a C credit necessary to achieve a desired level of abatement can be determined. This then determines the national cost of the policy, which can be compared with the sum of money the government is prepared to allocate to the policy.

3. Results and Discussion

The results of the analysis presented in table 2 have three main components. First, there is the distinction between changing enterprises from one of five main crops to either sheep or cattle, in two different farming systems: irrigated and dryland. Second, there is the difference between the cost of saving a tonne of CO_2-e to society and the cost to the farmer of saving CO_2-e per ha, for each of the two farming systems. Third, there is a distinction between the short-term annual costs and the long-term costs (over 25 years discounted using net present value (NPV) techniques at 7% per yr) in the two different farming systems. Note that the adjustment costs of making a change between enterprises and the compliance cost of a project are assumed to be zero.

The main findings to note are (a) for society, with the exception of wheat in the irrigated system, the net costs per t CO_2-e abated are between $462 and $1033 per yr. In the dryland system; with the exception of dryland cotton, the net costs per t CO_2-e abated are between $79 and $192 per yr. (b) For society as a whole, with the exception of irrigated wheat, in the long term these costs accumulate and in NPV terms amount to between $5444 and $11820 per t CO_2-e abated. In the dryland system, with the exception of cotton, the long term costs (in NPV terms) are between $1145 and $2458 per t of CO_2-e abated. (c) Farmers need an incentive to change enterprises and, with the exception of irrigated wheat, the net costs (i.e. loss of income) of changes in the irrigated system are between $847 and $1850 per ha in the first year. In the long term, the net costs increase substantially to between $9980 and $21670 per ha. In the dryland system, with the exception of cotton, the costs of changing enterprises is between $145 and $352 per ha in the first year. Over 25 yrs the net costs in the dryland system increase to between $2099 and $4507 per ha. (d) The negative results for irrigated wheat and dryland cotton are interpreted as the opposite to a cost; in other words these are the benefits of changing. A change from irrigated wheat to cattle is estimated to yield a benefit of $376 per t CO_2-e abated in one year and $4320 per t CO_2-e in the long term. Per ha, irrigated wheat farmers would improve earnings by an estimated $689 in the short term and $7920 over 25 yrs. The benefits are less appealing in a change from irrigated wheat to sheep. In the case of dryland, cotton producers changing to cattle is estimated to yield a benefit of $66 per t CO_2-e abated in one year and $550 per t CO_2-e over 25 yrs. Per ha, dryland cotton farmers would improve earnings by an estimated $121 in one year and $1008 over 25 yrs. The benefits are less appealing in a move from dryland cotton to sheep. (e) The question arises why farmers do not abandon irrigated wheat or dryland cotton and take up cattle? Regardless of the benefits of C storage, this would appear to be a profitable move at current commodity prices. However, a shortcoming of this analysis is that the costs of making the change have not been incorporated and these would need to be (in the case of irrigated wheat) less than $689 per ha in the first year and less than $7920 per ha over 25 yrs. Although the costs needed to make a similar change from dryland cotton to cattle, or from either irrigated wheat or dryland cotton to sheep are less, the same principle applies. In addition, while the losses in the long term would be less if a lower discount rate (say 4%) were used, any change from a cropping to a livestock enterprise (with the exception of those stated above) would still result in a loss. (f) Although changing from a cropping enterprise to sheep appears less cost-effective than a change to cattle, the adjustment cost of making this change may be less.

The government's draft Carbon Credits Determination Methodology 2014 identifies several factors that could change the net CO_2-e abatement for a change from a cropping to pasture [12], of which the most significant is the net negative effect of methane (CH_4) emissions from cattle or sheep grazing the pasture, which should be an overriding concern for society. The median values for these emissions for cattle [13] and sheep [14] are 83.5 and 7.3 kg CH_4/head/yr, respectively. Taking the global warming potential of CH_4 relative to CO_2 over 25 yrs as 63 [15], the effect of the CH_4 emissions from irrigated pasture is estimated to be 13.7 and 4.6 t CO_2-e/ha/yr for the cattle and sheep systems, respectively; for dryland pasture, the CH_4 effect would be 2.6 and 0.9 t CO_2-e/ha/yr. Notwithstanding the uncertainties in these estimates, the CH_4 emissions (in t CO_2-e/ha/yr) are much greater for irrigated pasture than the abatement of 1.835 t CO_2-e/ha/yr achieved through an increase in soil C under pasture. For dryland pastures, the situation is more favourable for sheep, which provide a net abatement benefit, but not for cattle.

Table 2. The net cost (per t CO_2-e and per ha) of changing from a cropping to a livestock enterprise to store soil C (assuming an average soil C storage of 0.5 t C/ha/yr).

Item	Units	Soybeans	Maize	Wheat	Lucerne	Cotton
Irrigation						
To society						
Immediate (1 yr)						
- to cattle	\$/t CO_2-e	508	462	-376	793	673
- to sheep	\$/t CO_2-e	748	702	-136	1033	913
Long term (25 yrs)						
- to cattle	\$/t CO_2-e	5977	5444	-4320	9302	7906
- to sheep	\$/t CO_2-e	8496	7962	-1802	11820	10425
To the farmer						
Immediate (1 yr)						
- to cattle	\$/ha	931	847	-689	1454	1234
- to sheep	\$/ha	1327	1243	-293	1850	1630
Long term (25 yrs)						
- to cattle	\$/ha	10959	9980	-7920	17054	14495
- to sheep	\$/ha	15576	14597	-3303	21670	19112
Dryland						
To society						
Immediate (1 yr)						
- to cattle	\$/t CO_2-e	81	79	178	109	-66
- to sheep	\$/t CO_2-e	95	93	192	123	-52
Long term (25 yrs)						
- to cattle	\$/t CO_2-e	1165	1145	2294	1493	-550
- to sheep	\$/t CO_2-e	1329	1309	2458	1657	-386
To the farmer						
Immediate (1 yr)						
- to cattle	\$/ha	148	145	326	200	-121
- to sheep	\$/ha	174	171	352	226	-96
Long term (25 yrs)						
- to cattle	\$/ha	2136	2099	4206	2737	-1008
- to sheep	\$/ha	2437	2399	4507	3038	-708

4. Conclusions

These results emphasize that enterprise choice is driven primarily by returns to the farmer. Thus, unless an incentive to change enterprises (such as a C credit) is large, change will not occur. This analysis, conducted over five different enterprises and two different farming systems, shows the best possible outcome because we have ignored the adjustment costs of any change, the compliance costs, and the intended 20% discount of a C credit for a 25-yr project. We have also assumed a generous value for an ACCU that is not likely to be achieved under the Direct Action Plan starting on 1 July 2015 (indeed the indicative cost of a C credit realized in the first auction held in April 2015 is \$13.95 per t CO_2-e). Hence all real situations present a worse-case scenario than that we have modelled here. With the exception of irrigated wheat and dryland cotton production, changing to grazed pasture to store soil C will not be a cost-effective way of abating GHG emissions. At a wheat price of \$200 per t, a change to cattle or sheep could be profitable provided the initial adjustment cost is less than \$689/ha or \$293/ha for a change to cattle or sheep, respectively. All the commodity prices considered here will fluctuate over time; in particular, wheat prices have been as high as \$300 per t which would make a change from wheat to cattle or sheep much less attractive than appears in our analysis.

Overall, the generally unfavourable cost-effectiveness of a change from cropping to pasture means that farmers are unlikely to make this change specifically to sequester soil C, but this general conclusion should not preclude individual farmers or farming groups from undertaking a more detailed analysis of the policy. However, irrespective of these financial results, the negative effect on net CO_2-e emissions of CH_4 emissions from cattle pastures (irrigated and dryland) and irrigated sheep pastures makes a change from cropping to grazed pastures an unsatisfactory option for abating GHG emissions nationally.

References

[1] Hunt G 2013 *Choosing the right market mechanisms for addressing environmental problems* http://www.greghunt.com.au/Portals/0/13-05-30%20Sydney%20Institute%20-%20choosing%20the%20right%20market%20mechanisms%20for%20addressing%20environmental%20problems.pdf accessed 25 October 2013

[2] www.environment.gov.au/climate-change/publications/emissions-reduction-fund-update (accessed 8 December 2014)

[3] Sanderman J, Farquharson R and Baldock J 2010 Soil carbon sequestration potential: *A review for Australian agriculture*. Department of Climate Change and Energy Efficiency, Australian Government Canberra

[4] Barnett A L, Schipper L A, Taylor A, Balks M R and Mudge P L 2014 Soil C and N contents in a paired survey of dairy and dry stock pastures in New Zealand *Agr Ecosyst Environt* **185** 34-40

[5] Condron L M, Hopkins D W, Gregorich E G, Black A and Wakelin S A 2014 Long-term irrigation effects on soil organic matter under temperate grazed pasture. *European Journal of Soil Science* doi: 10.1111/ejss.12164

[6] Anon. 2009 *Building soil organic carbon using biological farming systems in Australia's more intensive agricultural regions.* www.sba.asn.au/sba/pdf/BioCCS-build-soil-carbon.pdf accessed 21 August 2014

[7] www.environment.gov.au/climate-change/The Emissions Reduction Fund: Carbon Farming Initiative transition (accessed 8 December 2014)

[8] Levin H and McEwan P 2001 *Cost-effectiveness analysis: Methods and applications (2nd ed).* Thousand Oaks Sage

[9] Soussana J-F, Loiseau P, Vuichard N, Ceschia E, Balesdent J, Chevallier T and Arrouays D 2004 Carbon cycling and sequestration opportunities in temperate grasslands. *Soil Use Manag* **20** 219-30

[10] NSW Department of Primary Industries 2012 *Farm budgets and costs* www.dpi.nsw.gov.au/agriculture/farm-business/budgets (accessed 8 December 2014)

[11] Cowan L, Kaine G and Wright V 2012 The role of strategic and tactical flexibility in managing input variability on farms. *Syst Res Behav Sci* **30** 470–94, Wiley Online Library

[12] www.environment.gov.au/climate-change/emissions-reduction-fund/methods/sequestration-carbon-modelled-abatement-estimates (accessed 8 December 2014)

[13] Cottle D J, Nolan J V and Wiedemann S G 2011 Ruminant enteric methane mitigation: a review *Anim Prod Sci* **51** 491–51

[14] Hegarty R S, Alcock D, Robinson D L, Goopy J P and Vercoe P E 2010 Nutritional and flock management options to reduce methane output and methane per unit product from sheep enterprises *Anim Prod Sci* **50** 1026–33

[15] Forster P, Ramaswamy V *et al.* 2014 *Changes in atmospheric constituents and in radiative forcing* Climate Change 2007: The Physical Science Basis. Contribution of Working Group I

to the Fourth Assessment Report of the Intergovernmental Panel on Climate Change, chapter 2, table 2.14

IOP Conf. Series: Earth and Environmental Science **25** (2015) 012005 doi:10.1088/1755-1315/25/1/012005

Soil news – the soil carbon and climate policy journey in Australia and the role of different media

Ron Aggs
Climate and agricultural research and extension photo-journalist, PO Box 6114, Conder, ACT, 2906, Australia

E-mail: ron.aggs@gmail.com

Abstract. 'Enough soil carbon to mitigate climate change is a big ask' was a litmus piece in the October 2012 edition of Agriculture Today. The paper was the New South Wales Department of Primary Industries' flagship research, advisory and farm management vehicle, published monthly in The Land for 20 years, on the web since 2005 until December 2012. The October 2012 story dovetailed with Australian Broadcasting Corporation (ABC) television *Lateline* reporting that the Federal Coalition's (from now on Government's) climate policy could not demonstrate that storing carbon in Australian soils would achieve the major proportion of a target to reduce Australia's greenhouse emissions by five per cent on 2000 levels by 2020. It also provided background for the ABC's *FactCheck* verdict that voters in 2013 federal election were not hearing "the full story on climate research". The real story is how to inform urban Australia's poor understanding and lack of connection to how land managers must maintain and where possible improve soil quality for food security and food production as we adapt to climate change. And if you are in the business of information delivery or providing content, how do you choose your distribution channels to target as wide an audience as possible? One fundamental yardstick to avoid disenfranchising and discriminating against some people who want, and rely on, your information is to continually critically assess how fast high speed internet is reaching marginalised rural areas. Print is still the preferred news medium for the majority of farmers.

1. The 'Big Ask'

'Enough soil carbon to mitigate climate change is a big ask' was a definitive piece in the October 2012 edition of Agriculture Today, the New South Wales (NSW) Department of Primary Industries'(DPI) flagship research, advisory and farm management newspaper [1]. "Expectations were unrealistic for delivering increases in sequestration of carbon in soil", said one of Australia's most respected soil scientists Dr Mark Conyers, delivering the 2012 Harald Jensen lecture, hosted by the NSW Branch of the Australian Society of Soil Science. Dr Conyers is a DPI principal research scientist based at Wagga Wagga Agricultural Institute and the Harald Jensen lecture is a key annual event on Australia's soil science calendar. The content of Dr Conyers' story in *Agriculture Today* has never been contradicted to me, nor to my knowledge, to anyone associated with it.

The October 2012 story dovetailed with reporting by ABC TV's *Lateline* that the Federal Coalition's (now Australian Government's) 'Direct Action' climate policy, which included repealing Australia's carbon tax, could not demonstrate that storing carbon in Australian soils would achieve the major proportion of a target, widely criticised as too low, to reduce Australia's greenhouse emissions by five per cent of 2000 levels by 2020.

It also provided background for the ABC's *FactCheck* opinion that during the 2013 the federal election campaign, voters were not hearing "the full story on climate research" [2]. *Lateline* had twice reported the issue along similar lines in April 2011 [3]. Thereafter a concise and short timeline traced the disintegration of Direct Action's credibility: After a 2013 *Lateline* report [4] by Steve Cannane when host Tony Jones interviewed, now Environment Minister, Greg Hunt the Coalition went quiet about the potential for storing carbon in soils, then briefly flirted with re-afforestation as the mainstay of its Direct Action policy. After that the government generally avoided discussing both methods and fell back on saying they were confident of easily reaching emissions reduction targets. So the key question to Mr Hunt and the government remained unanswered: "Tell us exactly what methodologies you know will work to meet your five per cent target, rather than what you keep claiming you are 'confident' about" [3].

RenewEconomy website reported (July 21, 2014) global investment bank HSBC saying repeal of the carbon price the previous week would leave Australia's resource-intensive economy "even more vulnerable" as the world moves in [the] opposite direction and "...this will impact not just its energy-based commodity exports, but also other commodities such as agriculture". HSBC said Australia risked "carbon isolation" because of its "backward move" [5].

At the National Press Club in March 2013, Chief Scientist Professor Ian Chubb agreed with Opposition Leader Bill Shorten's proposition that "degradation of the climate change debate is the cautionary tale for what happens if we abandon the field to the conspiracy theorists and keyboard warriors, the social media trolls and the angry shouts of talkback radio..." [6].

Two weeks later in a televised climate debate with economist and former Labor government climate advisor Ross Garnaut on *Lateline*, the leader of the Palmer United Party (PUP), Clive Palmer, rejected the Intergovernmental Panel on Climate Change's latest findings and offered an offhand conspiracy theory: "well, I can get a group of scientists together... and pay them whatever I want to, and come up with any solution. That's what's been happening all over the world on a whole range of things" [7].

On July 7, 2014 at a live televised National Press Club lunch, it was suggested to him that "real science doesn't work that way... and that Australia's top agricultural food and fibre scientists, including climate scientists considered your comments a slap in the face."

In a rambling, almost entertaining response he then said he believed in climate change and explained his take on the unexpected and bizarre occasion three days earlier when he enlisted Al Gore, former US Vice President and Nobel Peace Prize winner for his work in climate change activism [8].

PUP's media release said Mr Gore had convinced him "to further consider the impact of greenhouse emissions on the planet and the need for a global solution" [9].

With their balance of power in the new Senate, PUP and other newly aligned Senators voted on July 17 to repeal the carbon tax.

Since December 2013 Mr Palmer and his staff had received sufficient background on the key science about the overstated potential for storing carbon in soil but had made no public reference to it in his political horse trading with the new government over its Direct Action plan.

He had previously called Direct Action a "token gesture", a waste of money and hopeless [10].

In August, he quipped on ABC TV's Q and A program that PUP had copied Labor's emissions trading scheme but with variations: its blueprint included a floor price of $0, to start when Australia's major trading partners formally initiated their own versions [11].

On October 30, 2014 PUP supported Direct Action into legislation.

Before all this and beyond the unfolding politics, in March 2014, there had been no disagreement among the international and Australian keynote speakers in an audience of soil science researchers at the Soil Change Matters workshop in Bendigo, that the government's Direct Action target could not be achieved mainly by sequestering carbon in soil.

For them the conundrum now was "don't throw the baby out with the bathwater".

While many thought Direct Action would fail if it was legislated, the Carbon Farming Initiative (CFI) was thought of as "something we need to make work", to enable carbon trading as part of any emissions reduction plan, and to facilitate the main game of improving land productivity and management, after farmers assessed associated financial and practical risks, including proof of permanence.

There are many issues to be resolved for the CFI to prove effective. The Federal Department of Environment maintained that 'approved sequestration methodologies' would still be required for agriculture and forestry. In early September 2014 the likelihood of who would pay the bill for greenhouse gas reduction had shifted onto taxpayers instead of polluters.

A methodology on soil carbon was still being evaluated. Whether the land sector had to compete with other sectors, such as stationary energy, or whether permits would be reserved for the land sector was unresolved. If the land sector has to compete it was unlikely to get much of a look in - especially with the potential for soil carbon being overstated.

The opportunities for forests to contribute were still very constrained by philosophical issues embedded in the current policy - largely, the peripheral issues around water and biodiversity imposed by the Greens when they held the balance of power with the Labor Party. For example, only providing credits for native forests if they were locked up and not actively managed - even though research shows that actively managed forests have a greater greenhouse gas reduction benefit, because the carbon in the harvested products is then locked up in processed timber products for more than 100 years, while the forest re-grows, sequestering more carbon [12].

Then there is the issue of forcing forest plantation owners to purchase high security water entitlements to cover the total amount of evapotranspiration from the plantation. This is despite other land uses (e.g. tropical pasture) having a greater impact on the water cycle [13] and not having to purchase entitlement. For example, the move to tropical pasture in northern NSW. While tropical pasture consumes less water per hectare than forests, it covers a much larger area. With 400,000 ha of temperate pasture replaced by species to suit more tropical conditions over the last 10 years [14], and tropical pasture using 0.1ML/ha/y more water than temperate pasture, this represented an impact that nobody was thinking about. It would be interesting to see if these policy prescriptions changed with the new Senate make up.

In the context of the Soil Change Matters workshop at Bendigo, an equally important story is to repeat the question that farmers, researchers, public and private sector advisory personnel and agribusiness people have already wrestled with for years, that is: how to improve urban Australia's poor understanding of and lack of connection to the ways land managers must maintain and where possible improve soil quality for food security and food production as we adapt to climate change?

With world populations exploding, what a substantial survival job that is, given some people's rhetoric that Australia could become "the food bowl of Asia". Federal Agriculture Minister Barnaby Joyce said it was not possible, and even doubling Australia's current food production would make a only a minor contribution to feeding Asia's growing middle class [15].

Leaving aside political rhetoric about climate policy, no-one in agricultural science, extension or production would debate the value of individual landowners doing whatever they could to cost-effectively store as much carbon as their soils permit, to enhance productivity.

2. How best to reach audiences with the latest news?

If you are in the business of 'information delivery' or 'providing content' - newspeak these days for 'news' - how do you choose your distribution channels to target as wide an audience as possible when you need to?

One fundamental yardstick, so as to avoid disenfranchising and discriminating against some people who want and rely on your information, is to continually critically assess how fast high speed internet is reaching marginalised rural areas. Otherwise, you may continue to reach major sections of agribusiness but risk leaving behind the most important group – the farmers. The highly complex mix of traditional and new media technologies and pathways now totally cater to metropolitan audiences and readers, and to regional urban centres almost as well. But citing NSW as the example, it would be a major blind spot to assume farmers (geographically more isolated outside urban centres) are migrating just as rapidly to predominantly consume web-based news and current affairs in preference to print [16].

The Audit Bureau of Circulations put sales of the print edition of *The Land* newspaper above 40,000 a week, which translated – weekly - to a readership of more than 100,000.

The most recent *Quantitative Agricultural Readership Survey* (QARS) by McNair Ingenuity for Fairfax Agricultural Media, demonstrated farmers were not yet migrating to the web or social media as their preferred sources of news [17]. Conducted every 2-3 years, QARS claims to be the most current and significant research on the habits and demographics of broadacre farmers in Australia. It investigates how technology is impacting on farmers, their consumption habits including how they consume media, and how they can be reached [17].

QARS' 2012 numbers suggest that anyone who really wants to keep contact with a mass farmer audience should maintain a presence in newspapers and other print formats if you have access, and if you don't, then initiate it, even for the short term. At least until there is clear market evidence that high speed internet access has convinced farmers to get their news online in preference to reading print; and currently that is a slower process than among metro and regional urban audiences. As it had in 2009, the 2012 QARS revealed the largest single farmer readership group of the print edition of *The Land* continued to be the 18-39 year olds – in 2013, 90 per cent of them read the paper and preferred print as the delivery medium. Of all farm management demographic groups, they would have been considered mostly like to mass-migrate online but at that stage, it had not happened [17]. In fact the print version of *The Land* is the preferred way farmers in all other older age groups surveyed also wanted their news presented [17].

Part of the argument supporting this niche market longevity of print to this point has been about farmers' lifestyles. They still regard print as the most convenient format to source their information - they can read before they go out to the paddocks early in the morning, again at lunchtime if they come in, or in the evening. Or take it with them in their vehicle. The other part of the argument is that access to high speed internet is limited in many rural areas.

So after 20 years of publication, why would the NSW Department of Primary Industries (NSW DPI) axe its widest reaching news vehicle, the flagship farm research, advisory and management newspaper, *Agriculture Today*? The paper had published monthly in the print edition of *The Land* as a highly credible masthead until December 2012 and on the web since 2005. It had approximately 105,000 readers a month. Asked specifically in February 2013 about its demise, Professor Robert Picard, Research Director, Reuters Institute for the Study of Journalism at Oxford University said:

> You don't want to walk away from your audience. So, I mean, any specialised publication is going to have a much stronger psychological bond between that publication and your provider and the audience than a general one. And that's always the case. So at this point if it's working, you know, if it ain't broke don't fix it. While people are out in the field they might want to check and be able to call back home. But they're not going to take their tablet out on their tractor. It just doesn't work that way. And you know, we have to think of technology as serving purposes and what they serve in people's lives. [18].

When 2011-13 budget cuts and rationalisation created one new organisation (Local Land Services) and restructured another (NSW DPI), *Agriculture Today* was delivering approximately 350 targeted research and advisory stories a year at a cost to the former NSW DPI of a few cents per message per reader. This cost was less than the cost to NSW DPI of an employee logging onto the internet each day before salaries of any media staff users or online content creators were factored in, including daily use of NSW DPI's Twitter account. It may seem a luxury for a government department to run its own newspaper but this is one portfolio – food and fibre research and extension – where it had been the perfect tool to reach core clients cost effectively, directly, for two decades. Now, at a time of the largest-ever reforms to the state's agriculture and natural resource management practices being implemented by the NSW government's new organisation, Local Land Services (LLS), the *LLS website* [19] was the go-to source – a massaged and visually manicured shopfront, heavy on enthusiastic public relations. LLS started operating in January 2014, a combination of three previous organisations – the extension (field advisory) staff of NSW Agriculture, the NSW Catchment Management Authorities and the NSW Livestock Health and Pest Authorities. Eight months in, its internet shop front talks a big game on the future of land and water management in NSW. Is it too early to predict the success and substance of the new enterprise?

On August 28 (2014) in his long-running Peppercorn column in *The Land*, Peter Austin wrote:

> From all accounts [LLS] is turning out to be just the cultural and administrative mish-mash many of us predicted. But at least under the previous structure everyone – constituents and staff – knew where they stood, and what was expected of them. Now we have a blancmange of an agency whose messages will necessarily be blurred... Its officers will find it hard to win the landholder trust that was vested in their forbears, whose roles were more clearly defined and understood [20].

In my opinion, the choices of media platforms to use to inform stakeholders about services in the split-up of NSW DPI and formation of LLS is a case study in hastily dispensing with one highly cost-effective, saturation coverage medium on the incorrect assumption that audiences could or would automatically, quickly convert to "new" (including social) media as a replacement.

Abolishing *Agriculture Today* was a cost cutting measure to save a relatively modest amount, pitched as a case of government showing leadership to farmers in adopting new media methods and technology and dispensing with what was supposedly outdated, rather than using them effectively in tandem. In the case of NSW DPI's remaining research, biosecurity and other programs', their main news vehicle then became the department's media release website [21].

Google Analytics in late February 2014 showed the entire DPI site had 331,000 hits in the previous month on all pages, without knowing how many hits there were on the media releases pages, which target journalists and news organisations rather than the department's wider client base. Regional print journalists may choose to run media releases or not for rural readers as prominently, or less prominently than *Agriculture Today* did, or with their own interpretation of each story.

NSW DPI's Twitter account at the end of May 2013 had 2000 followers, by March 23, 2014 it had 3040, by September 1, 2014: 3548 [22]. Not big numbers. There is something about government public relations messaging that stutters clumsily in Twitter's abbreviated conversation format.

A very telling print anecdote is that in a media release on the NSW DPI website, according to Dr Cameron Archer AO, Principal at Tocal College, sales of agriculture-related books were booming, defying the shift to digital e-reading. "Sales of books from Tocal have never been higher and are up significantly on last year," Dr Archer said. "In 2013 Tocal printed and distributed 13,000 publications Australia-wide and internationally on agriculture and related matters. The range of books includes more than 100 titles" [23].

Depending how long the National Broadband Network roll-out takes in the bush, farmers will still be likely to want news via print as a main source in the short term at least. Even though profits may decrease, The Land and corresponding interstate mastheads have the capacity to remain profitable, according to Fairfax, which anticipates the next QARS will update the picture in early 2015. In early September (2014), in a follow-up interview, Fairfax Agricultural Media national sales manager Ian Thomson repeated the organisation's assertion that in this niche market, specialist agriculture-focused printed newspapers will continue be profitable for at least 4-5 years.

References

[1] Aggs R 2012 Enough soil carbon to mitigate climate change is a big ask *Agriculture Today* [Internet] October [Accessed 2014 Aug 29]. Available from: www.dpi.nsw.gov.au/archive/agriculture-today-stories/ag-today-archive/october-2012/enough-soil-carbon-to-mitigate-climate-change-is-a-big-ask

[2] Soil carbon targets difficult to reach [television broadcast] *Lateline* Sydney: ABC; 2013 Apr 18. Available from: www.abc.net.au/lateline/content/2013/s3740395.htm
[3] Greg Hunt not giving the full story on climate research [television broadcast] *FactCheck* ABC, 2013 Aug 30. Available from: www.abc.net.au/news/2013-08-30/greg-hunt-not-giving-full-story-climate-research/4923258
[4] Coalition's direct action plan [television broadcast] *Lateline* ABC, 2011. Available from: www.abc.net.au/lateline/content/2013/s3740397.htm
[5] Parkinson G 2014 HSBC: Australia risks economic isolation from carbon repeal [Accessed 2014 Aug 20] In: *RenewEconomy* [Internet] Sydney. Available from: http://reneweconomy.com.au/2014/hsbc-australia-risks-economic-isolation-from-carbon-repeal-44643
[6] Chubb I 2014 Address to the National Press Club Canberra Mar 18 **33-34**
[7] Clive Palmer and Ross Garnaut debate carbon tax ahead of WA Senate election re-run [television broadcast] *Lateline* Sydney: ABC; 2014 Apr 4. Available from: http://www.abc.net.au/news/2014-04-04/clive-palmer-ross-garnaut-wa-senate-election-climate-carbon-tax/5366202
[8] Palmer C 2014 Address to the National Press Club Canberra Jul 7 **31-34**
[9] Proposed amendments to carbon tax repeal [Accessed 2014 July 4] In: Palmer United Party media release]. Available from: http://palmerunited.com/2014/07/proposed-amendments-carbon-tax-repeal
[10] Greg Hunt and Clive Palmer resist criticism of deal to pass Direct Action policy through Parliament [Accessed 2014 Oct 31] ABC website. Available from: http://www.abc.net.au/news/2014-10-30/hunt-palmer-resist-criticism-of-direct-action-deal/5852708
[11] Class, Cooling and Corruption [television broadcast] *QandA* program Sydney: ABC; 2014 Aug 18. Available from: http://www.abc.net.au/tv/qanda/txt/s4047130.htm
[12] Oliver M 2013 Do carbon offsets work? The role of forest management in greenhouse gas mitigation. *Science Findings*. 155, **1-7**. United States Department of Agriculture Forest Service. Available from: http://www.fs.fed.us/pnw/sciencef/scifi155.pdf
[13] Braumana K, Freyberg D, Daily G 2012. Potential evapotranspiration from forest and pasture in the tropics: A case study in Kona, Hawai'i *J Hydrol* **440-441** 52-61
[14] Tropical grass project gains hold. *Future Farm Industries CRC* [Internet] December 2013. [Accessed 2014 Aug 29]. Available from: www.futurefarmonline.com.au/LiteratureRetrieve.aspx?ID=163703
[15] Will Australia ever be the 'food bowl of Asia'? [television broadcast] *FactCheck* ABC, 2013 Aug 31. Available from:http://www.abc.net.au/news/2014-08-31/australia-not-the-food-bowl-of-asia/5680282
[16] Aggs R 2012 Preference is for print news *Agriculture Today* [Internet] December [Accessed 2014 Aug 29]. Available from: http://www.dpi.nsw.gov.au/archive/agriculture-today-stories/ag-today-archive/december-2012/preference-is-for-print-news
[17] *National Sales - Fairfax Agricultural Media* 2012 Quantitative Agricultural Readership Survey (QARS) [Internet]. Sydney NSW: Fairfax Agricultural Media. Available from: http://rpsales.cfmnet.predelegation.com/qars-research-1021.html
[18] Picard R 2013 Address to the National Press Club Canberra Feb 12 **22-25**
[19] *Local Land Services* 2014 [Internet] NSW: NSW Government. [Accessed 2014 Aug 25] Available from: www.lls.nsw.gov.au
[20] Austin P 2014 Peppercorn: Super idea just won't float *The Land* Sydney Aug 28. [Accessed 2014 Aug 29].
[21] *New South Wales (NSW) Department of Primary Industries* [Internet] [Accessed 2014 Aug 29] Available from: www.dpi.nsw.gov.au/aboutus/news/all

[22] *Twitter - New South Wales (NSW) Department of Primary Industries* [Internet] [Accessed 2014 Aug 29] Available from: https://twitter.com/nswdpi

[23] *NSW Department of Primary Industries* 2014 Agriculture-related books in big demand [Internet]. Sydney NSW: NSW Department of Primary Industries; Feb 13 [Accessed 2014 Aug 29]. Available from: www.dpi.nsw.gov.au/aboutus/news/all/2014/agriculture-books

IOP Conf. Series: Earth and Environmental Science **25** (2015) 012006 doi:10.1088/1755-1315/25/1/012006

Advances in Soil Biology: What does this mean for assessing soil change?

Helaina Black[1] and Pauline Mele[2]

[1] The James Hutton Institute, Craigiebuckler, Aberdeen, UK
[2] Department of Economic Development, Jobs, Transport and Resources, Vic AgriBio, 5 Ring Rd Bundoora, Victoria, Australia

E-mail: Helaina.Black@hutton.ac.uk

Abstract Our interests in soil change are moving away from soil properties and increasingly towards changes in the processes and functioning of soils. Soil organisms are fundamental to dynamics and change in soils through their fundamental role in soil processes [1]. However it is only with recent technical and theoretical advances that we have started to establish quantitative relationships between soil biology and soil change (c.f. [2]). It is this predictive understanding that will enable us to fully integrate soil biology into the effective monitoring and sustainable management of soils. This paper outlines some of the recent advances in soil biology and discusses their relevance to monitoring and management.

1. Introduction

We can now 'see' the world below our feet in high definition thanks to progressive advances in genetic, imaging and biochemical techniques that are coupled with use of ever more powerful computational techniques. This combination is shedding new light on our understanding of the complexities of the soil ecosystem and its contribution to wealth creation, recreation and environmental sustainability [3]. With the development of genetic tools, we can extract DNA and RNA directly from soil and have far less reliance on isolating or culturing soil organisms to identify or count them. The application of genetic tools to soils around the world has demonstrated that soil ecosystems are amongst the most diverse and biologically active in the entire world, as illustrated in table 1. Estimates of the number of individual bacterial cells in a gram of soil are upwards of 10 billion, more than the number of people on earth [4]. The field of soil biology is being revolutionised with the rate of discovery of new species at weekly for bacteria and monthly for fungi [5]. This rate of discovery is likely to continue as DNA approaches are applied to less well-characterised groups (e.g. protozoa, acari, collembola) and to the multitude of diverse and rare soils and habitats around the world. There are now maps of the soil bacterial taxa of many soils of the world and we know which taxa represent the most abundant and which typically represent the rarer forms (e.g. www.earthmicrobiome.com). This type of information is required for the other soil biota groups and is being undertaken through the Global Soil Biodiversity Initiative (http://www.globalsoilbiodiversity.org).

Our understanding of the diversity of soil biology is rapidly extending beyond this classical taxonomic approach and is increasingly focussed on the characterisation and quantification of the functional

Table 1. Summary of the biological diversity associated with the major soil-dwelling organisms.

Group/taxon (size)	Species described	Completely sequenced genomes	Species estimates
Viruses (nm)	1,832	3954[1]	>>4,000,000
Bacteria (μm)	11,082	4520[1]	>4,000,000
Archaea (μm)	453	272[1]	>>10,000
Fungi (μm-mm)	24,645	587[1]	1,500,000
Protozoa (μm-mm)	114,109	1[2]	200,000
Nematodes (μm-mm)	15,000	963[3]	20,000
Collembola (μm-mm)	6,500	0[4]	15,000
Acari (mm)	20,000	5[4]	80,000
Isoptera (mm)	2,600	0[4]	10,000
Oligochaeta (mm)	3,650	?	8,000

[1] as described in [5] [2] https://www.sanger.ac.uk/resources/downloads/protozoa [3]as described in www.nematodes.org, [4]as described in http://www.arthropodgenomes.org/wiki

diversity of soil organisms. This is not straightforward since one species can have multiple roles across different soil processes (e.g. both the production and oxidation of methane can be carried out by the same soil microbes) and many species or organisms can carry out the same process (e.g. degradation of complex organic compounds). This principle is known as 'functional redundancy' and it is this feature that gives soils an enormous capacity to resist, recover and adapt to change. The ability to define and ultimately predict the implications of a change in soil biodiversity for a loss or the improvement in soil processes still remains the "holy grail" in soil biological research. We need to unravel the idiosyncratic relationships between soil biodiversity and function as recently illustrated by Neilson et al [6]. This may require revision of the more classical approach to functional soil biodiversity, which has been heavily reliant on the allocation of conventional knowledge of taxa to the assignment of function. It will require a revision of the role of soil organisms in (bio) geochemical and (bio) physical processes, of the models that represent these processes (e.g. [7]) and how environmental change alters these dynamics (e.g. [8]). The UK NERC Soil Biodiversity and Function Programme, implemented on a Scottish upland farm, developed several methodological approaches to explore these relationships [1].

DNA-based methods continue to develop at a rapid pace to identify and estimate the role of soil organisms in multiple soil processes including the mineralisation of nutrients, the control of pathogens and the remediation of pollutants. DNA cloning techniques have enabled detailed studies on the distribution of important functional groups of soil microbes. In Scotland, these have been used to explore the occurrence and distribution of ammonia oxidising bacteria across different land uses and soils [9]. More intensive use of DNA approaches takes us into the territory of quantitative PCR (qPCR) [10] and microarray methods [11] which can show the presence, and estimate the abundance, of a whole range of genes associated with different processes of interest. The application of qPCR methods have been used in Australia to survey the distribution and abundance of genes associated with N-mineralisation and fixation across three major Victorian soil types demonstrates the utility of this approach [12]. Further technological advances in four areas - metagenomics [13, 14, 15], metatranscriptomics [16], metabolomics [17] and proteomics [18] - have opened up exciting new ways to assess the function of soil biology. These methods enable the simultaneous investigation of multiple processes performed by a particular soil biological population at fixed point in time and space. And now we have the ability to link genetic methods with isotopic tools to directly link organism to function (e.g. NanoSIP, [19]). The increasing challenge is synthesising and interpreting the immense amount of data that these techniques generate.

2. Soil biology – what's where and why?

The structure and activity of soil biological communities is reflected in their biogeography. A key challenge is to be able to predict why particular organisms and functional groups of organisms occur, or do not occur, in particular locations. This is the foundation of being able to utilise soil biology in monitoring or management. In the case of bacteria, it has been proposed that this biogeography is controlled primarily by the edaphic features of an ecosystem [20]. This means that everything is *not* everywhere but instead is defined by the physico-chemical features across a range of scales from the soil pore/aggregate (μm of) the landscape (km). In Australia, there is significant investment in measuring the soil microbial diversity associated with regional grain production systems (e.g. www.soilquality.org.au) and more broadly across the major terrestrial biomes (e.g. www.bioplatforms.com.au/special-initiatives/environment/soil-biodiversity). Preliminary data from Australian soil biodiversity surveys suggests strong correlations between soil type, regional rainfall distribution patterns and land-use. A regulator of soil community diversity is soil pH. For example, we see higher abundance of the bacterial phylum Acidobacteria in the low pH Ferrosols compared to the high pH Calcarosols of the Mallee [21]. This agrees with trends observed in soil biodiversity surveys elsewhere [20]. In the UK, there has been a similar investment in exploring the biogeography of soil biology. The Defra funded SQID project investigated the structure and function of soil organisms in typical land uses across the British mainland [22]. This study demonstrated the overwhelming influence of land use on the structure and activity of soil biota (other than soil bacteria) with an underlying influence of other factors including soil chemistry and plant community structure. This research has moved on to establishing characteristic soil biological communities for different land uses, to establish a monitoring framework, and in quantifying the impact of drivers of change such as nitrogen deposition from point sources (e.g. [23]) and management systems [24] as a precursor to defining indicators of change. Thus, with new techniques available for assessing soil biodiversity, questions related to the impacts of land use, management, pollution and climate change can be assessed. Given that biodiversity expresses a biogeography, it is becoming apparent that change in soil biodiversity needs to be measured and interpreted within a regional context to ensure that pro-active management, remedial or protective actions are relevant and hence effective. What these studies have demonstrated is that soil biology cannot be investigated in isolation of other soil properties, soil processes and wider environmental factors.

3. Why bother with soil biology?

With increasing emphasis on recognising and quantifying the value of soils to society [25], the "value" of soil biology has been considered in various ways. Ultimately, it is recognised that soil biology, with its vital role in key soil processes, is valuable to food, water and energy security. All are reliant upon the role of soil biology in nutrient release, pest and disease regulation and retention and in the stability / erodibility of soils, amongst other key ecosystem services. Soil biology also has a vital role in regulating our climate, as it is the key in the turnover of soil organic matter and release of greenhouse gases [26]. Our habitats that are valued for their conservation status often rely upon soil organisms (e.g. symbionts) for their characteristic flora and fauna while many important pharmaceutical products have been developed from soil derived biological compounds. However can we really put an economic value on soil biology? In 1997, Pimentel and co-authors estimated that the global value of soil biodiversity exceeded 1.5 trillion dollars. Although this figure demonstrates the importance of soil biology to a wide range of ecosystem services, the economic values are difficult to relate to management within specific regions or for particular land uses. At a national level, Glenk and colleagues [27] explored the value of soil biology to the Scottish economy. Ultimately, this study demonstrated that the ability to value the contribution of soil biology to food, water, energy, conservation and other services is severely limited by local scale data and defined relationships between soil biology and these services.

There are a few examples globally where we can truly appreciate and realise the value of soil biology, which can be illustrated by examples from Australia. Here, soil is viewed as a resource for sustaining and improving agricultural production and profitability. There are two major drivers for investment in Soil Biology. The first relates to the mounting concern regarding the continued degradation of the soil resource (State of the Environment and National Soils RDE Strategy) and the second the escalating input costs of production and the perception that soil biology can offset these. Investment in soil biology is largely industry based with $45M invested by the Grains Research and Development Corporation (GRDC) since 1992 [28]. This has served to focus research activity on applied outcomes related to demonstrating and quantifying the agronomic significance and developing management strategies (and products) that promote and enhance beneficial processes for productivity gains. Increasing energy costs that are driving up input (fertiliser N/P, lime, biocides) costs, and a general concern that soils are degrading further with increasing levels of salinity, acidity and disease and declining or low levels of carbon are considered by a growing stakeholder group to represent the major constraints to sustainable profitability [29, 30]. Grains Research and Development Corporation (GRDC) is the major funder of soil biology research in Australia investing more than $16M in two initiatives from 2002-2007 and 2009-2014. An economic evaluation based on judgements of the possible levels of adoption and benefits in areas relating to nutrient balance, disease control (including inoculants) and inoculants (yield increasing) indicate a net present value of $32 million and a benefit/cost ratio of the order of 4. The time lag from 2002 to the break-even point is of the order of 20 years. Sensitivity analyses aiming to capture the possible range of adoption and benefits indicated a range from break-even to a net present (2009) value of $120 million [31]. The time lapse between the investment and the return, coupled with the considerable effort in integrating information into a regional context represent the two major challenges for ongoing investment in this area.

4. Where are we at now and what are some of the remaining challenges?

There has been a revolution in knowledge regarding soil biodiversity with the development of molecular and imaging techniques. An analysis of the history of technology shows that technological change is exponential. So, at the present rate, in the context of soil biology, we are on the upslope and will experience something like 100's of years of progress in 10 year (http://www.kurzweilai.net, 2001). Keeping the speed of advances in mind, the critical features of studying and, ultimately, utilising soil biology must be that they have a focus (a question), a scale (a time frame and spatial context), a multi-disciplinary approach and a pathway to adoption that is meaningful to multiple end-users. It is crucial that soil biology is integrated fully with soil assessments to gain a holistic appreciation of soil change in both space and time.

Examples of some big (and difficult) questions are those posed by Handelsman and co-workers [32] for microbial communities which can be expanded to soil biology in general:

- How resilient are soil biological communities in the face of rapid global change?
- Can soil biological communities help to buffer and mediate key elemental cycles which are now undergoing rapid shifts?
- Can changes in soil biological communities serve as sensors and early-alarm systems of environmental change?
- To what extent can we manage soil biological communities to modulate the effects of human activities on natural elemental cycles sensibly and deliberately?

We are advancing slowly towards some answers with massive genetic datasets being generated and analysed that will fill some of the knowledge gaps in defining species and function of soil biology across the soils of the world. There are also many outstanding challenges. A significant technical challenge is requirement for sophisticated informatics approaches that can process enormous amounts of data and statistical techniques to explore the non-linearity in relationships between soil biology and its environment. Similarly a significant social challenge is transforming the way we think about and investigate soil biology. This requires a greater attention to systems and non-linear thinking. In practical terms, this means increasing use of metadata, often described as the 'halo of data', which includes soil chemical, physical and historical agronomic/land-use data that enables an accurate description of the soil as a habitat for biological functions. This, in turn, requires cooperation across a range of skill-sets from the soil biologist, soil scientists to experiential skills of land-holders. A noteworthy model of such a coordinated systems-based approach is provided by the human microbiome project [33] which shares a central tenet with soil biodiversity; the absolute reliance on a diverse microbial community to sustain a healthy system, whether a human body or a soil. No doubt these are exciting times for mapping the biogeography of soil biology and the changes in functionality as a result of human interventions. This will not only challenge some long held ecological concepts borrowed from above-ground systems but will generate indisputable evidence of soil biology as an asset worth protecting.

References

[1] Bardgett RD, Usher MB and Hopkins DW 2005 Biological diversity and function in soils. Cambridge University Press.
[2] Wall DH and Nielsen UN 2012 Biodiversity and Ecosystem Services: Is it the Same Below Ground? *Nature Education Knowledge* **3**(12), 8
[3] Ritz K, Black HIJ, Campbell CD, Harris JA and Wood C 2009 Selecting biological indicators for monitoring soils: a framework for balancing scientific and technical opinion to assist policy development. *Ecol Indic* **9**, 1212–21
[4] Nature Reviews Microbiology, Editorial **9**, 628 (September 2011) doi:10.1038/nrmicro2644
[5] NCBI 2014 data from www.ncbi.nlm.nih.gov
[6] Neilson UN, Ayers E, Wall DH and Bardgett RD 2011 Soil biodiversity and carbon cycling: a review and synthesis of studies examining diversity-function relationships. *Euro J Soil Sci* **62** 105-16
[7] Bradford MA, Tordoff GM, Black HIJ, Cook R, Eggers T, Garnett MH, Grayston SJ, Hutcheson KA, Ineson P, Newington JE, Ostle N, Sleep D, Stott A and Jones TH 2007 Carbon dynamics in model grassland with functionally different soil communities *Funct Ecol* **21** 690-97
[8] Heath J, Ayres E, Possell M, Bardgett RD, Black HIJ, Grant H, Ineson P and Kerstiens G 2005 Rising atmospheric CO_2 reduces sequestration of root-derived soil carbon. *Science* **309** (5741), 1711-13
[9] Yao H, Campbell CD, Chapman SJ, Freitag TE, Nicol GW and Singh BK 2013 Multi-factorial drivers of ammonia oxidizer communities: evidence from a national soil survey. *Environ Microbiol* **15** (9), 1462-2920
[10] Sharma S, Radl V, Hai B, Kloos K, Fuka MM, Engel M, Schauss K and Schloter M 2007 Quantification of functional genes from procaryotes in soil by PCR *J Microbiol Meth* **68** 445–52
[11] He Z, Gentry TJ, Schadt CW, Wu L, Liebich J, Chong SC, Huang Z, Wu W, Gu B, Jardine P, Criddle C and Zhou, J 2007 GeoChip: a comprehensive microarray for investigating biogeochemical, ecological and environmental processes. *The ISME Journal* **1**, 67–77; doi:10.1038/ismej.2007.2
[12] Hayden HL, Drake J, Imhof M, Oxley AP, Norng S, and Mele PM (2010). The abundance of nitrogen cycle genes (*amoA* and *nifH*) depends on land-use and soil type in South-Eastern Australia. *Soil Biol Biochem* **42**, 1774-83
[13] Daniel R 2005 The metagenomics of soil *Nat Rev* **3** 470-78
[14] Delmont TO, Robe P, Cecillon S, Clark IM, Constancias F, Simonet P, Hirsch PR, Vogel TM 2011 Accessing the Soil Metagenome for Studies of Soil Diversity *Appl Environ Microb* **77**

(4) 1315-24
[15] Wooley JC, Godzik A and Friedberg I 2010 A Primer on Metagenomics. *PLoS Computational Biol* **6** (2) 1-13
[16] Carvalhais LC, Dennis PG, Tyson GW and Schenk PM 2012 Application of metatranscriptomics to soil environments *J Microbiol Meth* **9** (2) 246-51
[17] Simpson MJ and McKelvie JR 2009 *Anal Bioanal Chem.* **394**, 137-149
[18] Giagnoni L, Magherini F, Landi L, Taghavi S, Modesti A, Bini L, Nannipieri P, Van der Lelie D. and Renella G 2011 Extraction of microbial proteome from soil: potential and limitations assessed through a model study. *Euro J Soil Sci* **62** 74-81
[19] Pett-Ridge J and Weber PK 2011 NanoSIP: NanoSIMS applications for microbial biology. Ed Navid A. Microb Microbial Systems Biology: Methods and Protocols. Methods in Molecular Biology 881. Springer
[20] Fierer N and Jackson RB 2006 The diversity and biogeography of soil bacterial communities. *PNAS* **103** (3), 626-31
[21] Mele P, Hayden H, Methe B, Stockwell T, Pfannkoch C, Lewis M, Tanenbaum D, Rusch D and Heidelberg K 2008 Assessing soil microbial communities using a metagenomic approach. 12th International Symposium on Microbial Ecology, Cairns Australia.
[22] Black HIJ, Ritz K, Harris JA, Cameron CM, Campbell CD, Chamberlain PM, Creamer R, Pawlett M, Wood C and Singh BK 2011 Scoping Biological Indicators of Soil Quality Phase II. Defra Final Contract Report SP0534. (http://randd.defra.gov.uk)
[23] Mitchell RJ, Beesley L, Briggs R, Cuthbert A, Dawson JJC, Helliwell RC, Hewison RL, Kerr C, Leith ID, Owen IJ, Potts JM, Newman G, Smith D, Shephard LJ, Sturgeon F, Taylor AFS, White D, Williams A, Williams E and Black HIJ 2013 Testing soil quality indicators. Scottish Environment Protection Agency Commissioned Report No.HP1108.
[24] Vink SN, Neilson R., Robinson D and Daniell TJ 2014 Temporal and land use effects on soil bacterial community structure of the machair, an EU Habitats Directive Annex I low-input agricultural system *Appl Soil Ecol* **73** 116-23
[25] Costanza R, d'Arge R, de Groot R, Farber S, Grasso M, Hannon B, Limburg K, Naeem S, O'Neill RV, Pareulo J, Raskin RG, Sutton P and van den Belt, M 1997 The value of the world's ecosystem services and natural capital. *Nature* **387**, 253-26
[26] Mele PM 2011 Soil health, soil biota and climate change. Chapter 8. In 'Soil Health and Climate Change'. Editors: Bhupinder P Singh, Annette L Cowie & K Yin Chan. Soil Biology Series (Springer, Amsterdam).
[27] Glenk K, McVittie A, Towers W, Watson C and Black HIJ 2010 Socio-economic data on Scottish soils – collection and development. Scottish Environment Protection Agency Commissioned Report. [online] www.sepa.org.uk/land/idoc.ashx?docid=ebebd86e-4af1-4a48-b2e4-952c534fce8e&version=-1
[28] Mele PM 2009 A New Charter for Investing in Soil Biology RDE in Australia. *Occasional Publication*. Land and Water Australia. Canberra.
[29] Campbell A 2008 Managing Australia's Soils: A Policy Discussion paper. Canberra 83.
[30] Sanderman J, Farquharson R and Baldock J 2010 Soil Carbon Sequestration Potential: A review for Australian agriculture CSIRO Land and Water Canberra.
[31] Agtrans Research 2009 Impact Assessment: An Economic Analysis of Investment in the GRDC Soil Biology Program (Themes 1-3) pp 1-35
[32] Handelsman J , Tiedje J, Alvarez-Cohen L, Ashburner M, Cann IK, DeLong EF, Doolittle WF, Fraser-Liggett CM, Godzik A, Gordon JL, Riley M and Schmid MB 2007 The New Science of Metagenomics: Revealing the Secrets of our microbial planet. The National Academies Press Washington, DC.
[33] Turnbaugh PJ, Ley RE and Gordon JI 2007 The human microbiome project: exploring the microbial part of ourselves in a changing world. *Nature* **18** 804-10

The Soil and its Chemistry- Critical Futures

Mike J. McLaughlin
CSIRO Land and Water/University of Adelaide, Waite Campus, PMB 2, Glen Osmond, SA 5064, Australia

E-mail: Mike.Mclaughlin@csiro.au

Abstract. Unlike changes over time to the earth's atmosphere or oceans, changes to soils occur at a very local scale, with much less interconnectedness between the points of change. A further difference between soils and other media is the much greater buffer capacity of the system to change, which means that changes are more gradual and difficult to measure. Further difficulties of assessing changes in soil "condition" are due to the natural heterogeneity of soils across landscapes, which poses sampling, analysis and statistical interpretation challenges of a much greater magnitude than those for the atmosphere or for surface or ground waters. Superimpose localised anthropogenic disturbances and enhancements to soil on this (at a farm or paddock level) and it is evident that knowledge and assessment of soil "condition" is needed at a relatively small scale compared to assessments of atmospheric or aquatic system "condition". At the same time, soils contribute significantly at larger scales in modifying (in both good and bad ways) to the quality of water systems, to emissions of gases (greenhouse and others) to the atmosphere, and to feeding a steadily growing human population around the globe – so changes to soil really do matter. This juxtaposition of scale issues means that to understand the global impact of changes to soil "condition", we need to measure changes at small scales and integrate the mosaic of effects up to larger scales i.e. think globally, measure locally.

1. Rates of soil change

Changes to the chemistry of our soils have accelerated rapidly in the last 250 years due to the "industrialisation" of agriculture where inputs of chemicals and fossil fuel energy (in the form of mechanised equipment) has allowed deforestation, cultivation, fertilisation and farm chemical inputs on a much larger scale than observed in previous millennia – a good example for changes in soil organic carbon content over time was provided recently by Richter and Yaalon [1] (Figure 1).

Similarly here in Australia, we have seen dramatic changes over time in the chemistry of our soils due to the introduction of phosphatic fertilisers and legume-based pastures [2] (Figure 2) – and note these changes in soil properties have been predominantly positive in contrast to some negative yet beliefs about fertiliser use and soil health not backed up by good scientific data. Nonetheless, changes to soil chemical properties are generally slow gradual changes, due to the inherent inertia of the soil to additions or removal of elements, or to changes in composition in terms of minerals and organic matter – what soil scientists usually term the "buffer capacity".

Because of the slow nature of change, it is imperative that measurement technologies are both accurate and precise (discussed further in the presentation), and that we have "sentinel" sites that alert us to adverse changes – in soil science the sentinels are often long-term field experiments (e.g. Rothamsted, Morrow plots, Waite Long Term Rotation Experiment) that provide invaluable information on how management practices affect crop production, soil properties, water quality and emissions of gases to the atmosphere [3].

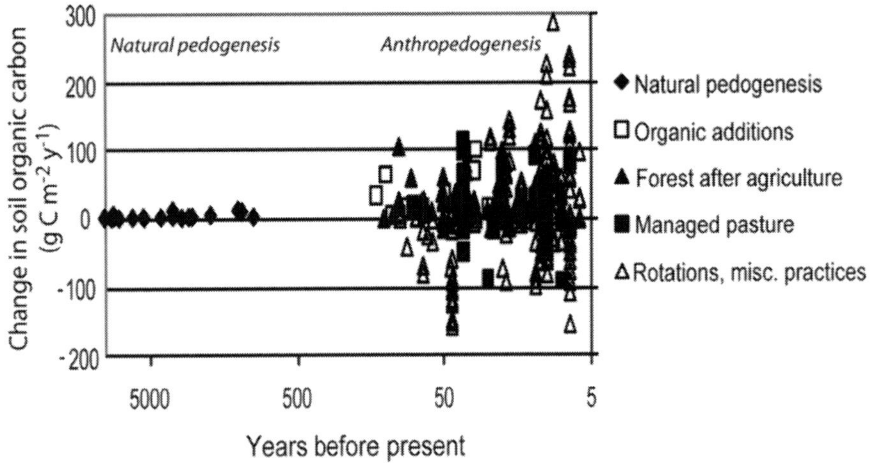

Figure 1. Rates of change in soil organic carbon in response to cultivation, manure amendments, reforestation, and other practices contrast with rates of change in soils that are products of natural soil formation. Plotted are average rates of change over the indicated age of soil or soil-management regime (from [1]).

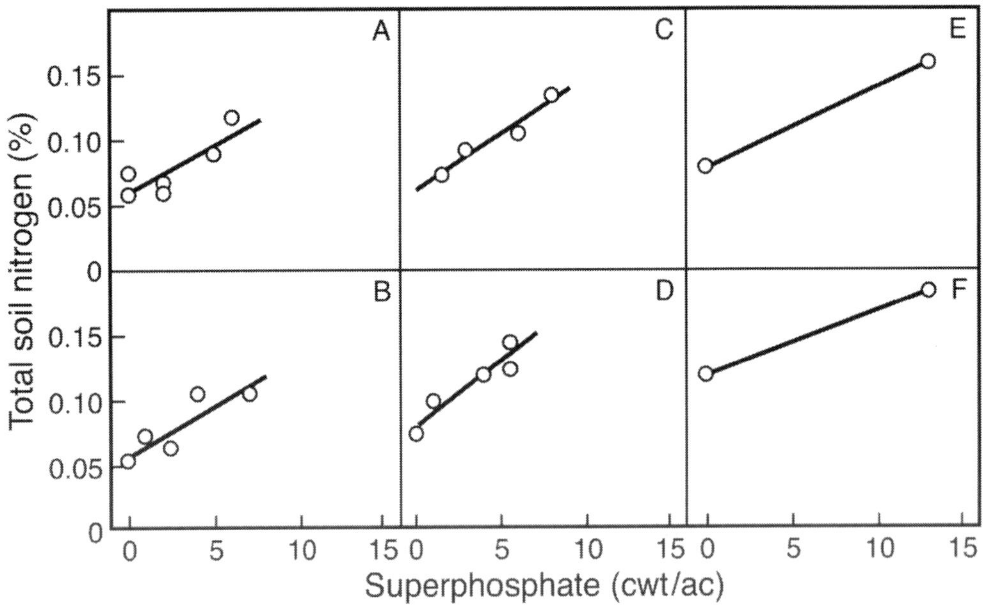

Figure 2. Relationship between amounts of single superphosphate applied to six properties and soil concentrations of nitrogen (from [2]).

2. Scale of changes in soil properties

As noted in Figure 1, use of soils in agricultural systems has in many cases increased the heterogeneity of soil properties across the landscape, and inference of soil condition from the pedogenic background state is becoming less relevant with time. In addition, pollution of soils from industrial or urban activities is also extremely localised, so that assessment of soil contamination (and remediation) is often confounded by variability of properties and analytes across quite small scales.

Traditional assessments of soil chemical "condition" have usually been by field sampling (at various intensities dependent more often on factors such as time, cost and enthusiasm rather than actual heterogeneity) and laboratory analysis (with the suite of analytes again more often dependent on time, cost and enthusiasm than anything else). Cost is usually the critical factor which drives the adoption and use of soil analysis, so that multi-analyte tests are now much preferred in many countries e.g. the Mehlich soil test widely used in USA [4]. Unfortunately the dominance of the cost driver in soil analysis means that the adoption of multi-analyte tests is often driven more by laboratory expediency than by predictive power.

New technologies have emerged in the last 20 years which have revolutionised assessment of soil condition, which allow much faster analysis of soils at much lower cost, and which can be placed in the hands of land managers – direct spectroscopic methods which are field portable. Good current examples of these are near- and mid-infrared soil spectrometers, as well as laser induced breakdown spectrometers, portable x-ray fluorescence spectrometers and laser-induced fluorescence instrumentation. It is important these powerful multi-analyte techniques are well calibrated and validated before adoption but I believe rapid, direct and inexpensive spectroscopic analysis of soils is key to assessing and managing soils at the local scale, where all the "action" is.

3. Conclusions

Changes in soil chemical properties will increasingly need to be measured at high spatial resolution by simple, low cost field-based techniques so that landscape-scale changes over time can be determined by an aggregation of data, or a using a spatial covariate, rather than compositing of soil samples over an area (to reduce analytical costs) and measuring by laboratory techniques.

References

[1] Richter, D and Yaalon, D 2012 "The Changing Model of Soil" revisited. *Soil Science Society of America Journal* **76** 766-778

[2] Williams, C and Donald, C 1957 Changes in organic matter and pH in a podzolic soil as influenced by subterranean clover and superphosphate, *Aust J Ag Res* **8** 179-89

[3] Korschens, M 2006 The importance of long-term field experiments for soil science and environmental research - a review *Plant Soil Environ* **52** 1-8

[4] Mehlich, A 1984 Mehlich-3 soil test extractant - a modification of Mehlich-2 extractant, *Commun Soil Sci Plan* **15** 1409-16

Key soil functional properties affected by soil organic matter – evidence from published literature

Brian Murphy

Honorary Scientific Fellow, NSW Office of Environment and Heritage, Cowra, NSW, Australia. In association with the Commonwealth of Australia (Department of the Environment) and the Australian Grains Research and Development Corporation

E-mail: brian.murphy@environment.nsw.gov.au

Abstract. The effect of varying the amount of soil organic matter on a range of individual soil properties was investigated using a literature search of published information largely from Australia, but also included relevant information from overseas. Based on published pedotransfer functions, soil organic matter was shown to increase plant available water by 2 to 3 mm per 10 cm for each 1% increase in soil organic carbon, with the largest increases being associated with sandy soils. Aggregate stability increased with increasing soil organic carbon, with aggregate stability decreasing rapidly when soil organic carbon fell below 1.2 to 1.5 5%. Soil compactibility, friability and soil erodibility were favourably improved by increasing the levels of soil organic carbon. Nutrient cycling was a major function of soil organic matter. Substantial amounts of N, P and S become available to plants when the soil organic matter is mineralised. Soil organic matter also provides a food source for the microorganisms involved in the nutrient cycling of N, P, S and K. In soils with lower clay contents, and less active clays such as kaolinites, soil organic matter can supply a significant amount of the cation exchange capacity and buffering capacity against acidification. Soil organic matter can have a cation exchange capacity of 172 to 297 cmol(+)/kg. As the cation exchange capacity of soil organic matter varies with pH, the effectiveness of soil organic matter to contribute to cation exchange capacity below pH 5.5 is often minimal. Overall soil organic matter has the potential to affect a range of functional soil properties.

1. Introduction

Much of the recent interest in soil organic matter has been as a vehicle to increase soil carbon levels in order to trade carbon or to gain benefits in market based instruments for carbon. However, it can be argued that the potential effects of soil organic matter on the productive capacity of soils are also of practical and economic importance and of significant interest to many in the agricultural community. It is the objective of this review, to concentrate on the capacity of soil organic matter to be an agent to maintain and improve soil condition and soil health and so the productive capacity of the soil.

While several reviews have been undertaken and suggest it is difficult to find quantitative evidence that soil organic matter improves soil properties [1], the approach taken here has been to investigate more directly the effects of soil organic matter on individual soil properties, especially through any pedotransfer functions that relate soil organic matter to the values of individual soil properties. It is clear that soil organic matter affects a range of soil properties and therefore any assessment of the influence of soil organic matter on productivity needs to consider this broad range of effects. In a final conclusion a paper that evaluates the overall effect of soil organic matter on overall productivity is reviewed. The detailed review is presented in [2].

2. Soil organic matter and functional soil properties

A summary of the effects of soil organic matter on some of the major functional soil properties based on the review is given below. The summary is based on an evaluation of the published information and pedotransfer functions and relationships relating the soil properties to soil organic matter levels.

2.1. Water holding capacity

A range of pedotransfer functions that included soil organic matter as one of the variables were evaluated [3], [4]. Several features of the relationship between soil organic matter and soil water holding capacity became evident.

- As a general guide the plant available water between 10kPa and 1500kPa increased more for sandy soils with increasing soil organic carbon than for clayey soils. The general trend is shown below but this is only a very general guide, with more detailed data required for making specific recommendations:
  - Sandy soils by about 3 mm/100mm of soil for every increase of 1% soil organic carbon
  - Loam soils about 2.5 mm/100mm of soil for every increase of 1% soil organic carbon
  - Clayey soils about 2 mm/100 mm for every increase of 1% soil organic carbon.
- The amount of water held at 10 kPa and at 1500 kPa increased with increasing soil organic matter although the amount of water held at 10kPa increased at a greater rate.
- The increases in soil organic matter tend to be limited to the surface soils and probably the top 5 to 10 cm. This limits the capacity of increases in soil organic matter to increase the overall water holding capacity of a soil profile.
- The changes in water holding capacity associated with increasing soil organic matter are complicated by the effects of soil organic matter on bulk density.
- The predictions of the effects of soil organic matter on water holding capacity varied within the published literature and this is a reflection of the different data set and soils used to derive the different pedotransfer functions.

2.2. Aggregate stability

While not a functional property itself, aggregate stability is a fundamental physical property of the soil that influences many other soil physical properties. Good aggregate stability is generally considered to be required to maintain good soil structure and a suitable soil physical condition of the soil for plant growth, infiltration and control of erosion. In general a level of soil organic carbon of 2 to 2.5% is considered necessary to maintain good aggregate stability [5] and aggregate stability is considered to deteriorate rapidly when SOC falls below 1.2 to 1.5% [5]. The relative importance of soil organic matter in maintaining aggregate stability varies with texture. In sandy soils soil organic matter is the most important factor, in clayey soils cation balance can be the most important effect and in loamy soils both cation balance and soil organic matter are important [6].

2.3. Compaction characteristics and friability

Several published relationships showed that the compaction characteristics of soils, friability and the Atterberg Limits (strong indicators of the compaction and tillage characteristics of soils) are all strongly affected by soil organic matter and from an agricultural and cropping viewpoint are improved by increasing soil organic matter [7]. In general, When the soil organic carbon % falls below 1%, the compaction and friability characteristics of soils became more limiting for plant growth and tillage operations[8,9], but this requires further investigation. Again texture is important with more clayey soils less influenced by soil organic matter [7].

2.4. *Soil erodibility*

Soil erodibility is only one factor in determining the potential for water erosion with rainfall erosivity, length and degree of slope and land management factors such as cover and the presence of loosely tilled soil being other factors. However it when soil organic matter falls below 2% (SOC < 1.2%) soil erodibility is considered to increase [10]. Good aggregate stability in reducing erosion and so this is related to soil organic matter and especially the amount of aggregates > 125μ [11].

2.5. *Nutrient Cycling*

A major feature of soil organic matter is that it holds a relatively constant ratio of the different nutrients [12]. For every 1 tonne of soil organic carbon there is 83 kg of N, 20 kg of P and 14 kg of S [13]. This does vary between soils and with history of crops, pastures and land management, for example the amounts suggested by Williams and Donald [14] for long term pastures are 65 kg of N, 4.4 kg of P and 9 kg of S. In the decomposition of the soil organic matter these nutrients can be released into the soil although how much becomes available to plants depends on a range of factors. In the formation and then in the decomposition of soil organic matter, a flux of nutrients or a recycling of nutrients is an outcome. Thus soil organic matter can be an important sink and source of nutrients in agricultural production.

Nitrogen is a dynamic nutrient being continuously recycled between the atmosphere, the soil solution, soil organic matter, plant material and soil organisms. Soil organisms are responsible for the transformation of nitrogen between these different pools. Mineralisation of soil organic matter by which nitrogen in soil organisms, plant material and soil organic matter is converted to mineral N (nitrate and ammonium ions) and made available to plants is a major process for the provision of N. Effectiveness of mineralisation depends on the chemical characteristics of the substrate material particularly its C:N ratio and its lignin content. Substrates with high C:N ratios (generally cereals) are likely to result in N being fixed in the soil organism pool and becoming unavailable to plants. While the mineralisation of soil organic matter under aerobic conditions results in the formation of nitrate and ammonium ions, under anaerobic conditions the gases N_20 and N_2 can be formed rather than mineral N.

Phosphorus occurs in several pools in the soil and in many soils the P from soil organic matter may be only one of these pools (Shen et al.)[15]. It estimated that P from soil organic matter is generally about 40% of total P in the soil, but this can vary from 20% to 80% depending on the soil type. Soil organic matter can prevent P being fixed into unavailable forms by the Fe and Al minerals [16] keeping it in a form which remains in the available pool for plants, even if this organic P needs to be mineralised before it becomes immediately available to plants. The mineralisation of plants and soil organic matter can be a major source of P for plants [17].

Sulfur is predominately supplied from the soil organic matter in most soils although some soils have a high mineral level of S such as those high in gypsum and some volcanic soils and some soils associated with marine deposits having been associated with acid sulphate soils.

The soil microbial population can be a critical factor in the effectiveness of nutrient cycling through soil organic matter. Maintaining soil biological health is an important factor in the functioning of soil organic matter [18]. Land management can influence the microbial populations in soils and their effectiveness in nutrient cycling to maximise the amount of nutrients that become available to plants [18], [19].

2.6. *Cation exchange capacity*

It is clear from the published information that soil organic matter can have an effect on cation exchange capacity but it is complex as it is dependent on the texture of the soil, but also on the pH range of the soil [20]. Much of the soil organic matter fraction that contributes to the cation exchange capacity has variable charge and this is why the effect of soil organic matter on CEC is pH dependent. Below pH 5.5, it appears that soil organic matter does not contribute greatly to CEC. The relationships between soil organic matter and CEC were shown for sets of well-defined soils [21], [22] but the relationships were different for the different soil groups. For a chernozem group of soils the CEC of the soil organic matter was 297 cmol(+)/kg and 134 for a group of acid soils [21]. In red earths, it was estimated the soil organic matter had a CEC of 172 cmol(+)/kg [22]. It would appear that most of the charge contributing to the CEC is from the humus fraction.

2.7. *Buffering capacity to acidification*

The capacity of soil organic matter to buffer the soil against soil acidification has been recognised in both Queensland and NSW [23], [24]. Its capacity to buffer against acidification is also dependent on texture and clay content. While increasing the soil organic matter can increase the buffering capacity against soil acidification (possibly up to 3 tonnes of calcium carbonate by increasing soil carbon by 1 to 2%), in general this will only provide short term slowing of acidification as the acidification rates will generally be sufficient to overcome the increased buffering by higher soil organic matter levels, up to 10 to 15 years based on Fenton and Helyar [25]. However, this increased buffering capacity may provide short term agronomic and economic benefits in some soils.

3. Assessing the overall impact of soil organic matter on functional soil properties

Soil organic matter can affect a wide range of soil physical and chemical properties. It can affect the nutrient cycling and nutrient availability in soils. Predicting the overall impact of soil organic matter on productivity is complex because of the wide range of potential effects of soil organic matter. The effects are also likely to be dependent on seasonal conditions and soil type, especially soil texture. Indications are that once soil organic carbon levels fall below 1%, then a wide range of soil properties can be adversely affected. Given the complexity of the potential effects of soil organic matter on soil functions, the use of biophysical models such as ASRIS, is suggested as a useful method to examine the overall effects of soil organic matter on productivity in more detail.

References

[1] Loveland, P and Webb, J. 2003 Is there a critical level of organic matter in the agricultural soils of temperate regions: a review *Soil Till Res* **70** 1 – 18.

[2] Murphy, B, W, 2014 *Soil organic matter and soil function – Review of the literature and underlying data.* Department of the Environment, Canberra, Australia. http://www.environment.gov.au/climate-change/publications//soil-organic-matter-soil-function

[3] Williams, J, Ross, P and Bristow, K. 1992 *Prediction of the Campbell water retention function from texture, structure and organic matter.* In M. Th. Van Genuchten and FJ Leij (eds). *Indirect methods for estimating hydraulic properties of unsaturated soils.* Proceedings of International Workshop on Indirect Methods for Estimating the Hydraulic Properties of Unsaturated Soils. Riverside, California, October 11 – 13, 1989. US salinity Laboratory, Agricultural research Service, US Department of Agriculture, Riverside, California.

[4] Rawls, WJ, Ahuja, LR and Brakensiek, DL 1992 *Estimating soil hydraulic properties from soils data.* In M. Th. Van Genuchten and FJ Leij (eds). *Indirect methods for estimating hydraulic properties of unsaturated soils.* Proceedings of International Workshop on Indirect Methods for Estimating the Hydraulic Properties of Unsaturated Soils. Riverside, California, October 11 – 13, 1989. US salinity Laboratory, Agricultural research Service, US Department of Agriculture, Riverside, California.

[5] Kay BD and Angers, DA 1999 *Soil Structure.* In '*Handbook of Soil Science*'. (Ed ME Sumner). ppA229 – A276. (CRC Press:Boca Raton, USA).More references

[6] Oades, JM. 1993 The role of soil biology in the formation, stabilization and degradation of soil structure. *Geoderma.* **56**, 377-400.

[7] Keller, T and Dexter, A 2012 Plastic limits of agricultural soils as functions of soil texture and organic matter *Soil Res* **50**, 7 – 17.

[8] Macks, SP, Murphy, BW, Cresswell, HP and Koen, TB. 1996 Soil friability in relation to management history and suitability for direct drilling *Aust J Soil Res* **34**, 343 – 60.

[9] Thomas, GW, Hazler, GR and Blevins, RL. 1996 The effects of organic matter and tillage on maximum compaction of soils using the Proctor test. *Soil Sci* **161**, 502 – 08.

[10] Rosewell, CJ and Loch, RJ 2002 *Estimation of the RUSLE soil erodibility factor.* In Neil McKenzie, Kep Coughlan and Hamish Cresswell (eds). *Soil Physical Measurement and Interpretation for Land Evaluation.* CSIRO Publishing, Collingwood Australia.

[11] Loch, RJ and Foley, JL 1994 Measurement of aggregate breakdown under rain. Comparison with tests of water stability and relationship with field measurement of infiltration. *Aust J Soil Res* **32** (4). 701 – 720.

[12] Kirkby, CA, Kirkegaard, Richardson, AE, Wade, LJ, Blanchard, C and Batten, G 2011 Stable soil organic matter: A comparison of C:N:P:S ratios in Australian and other world soils. *Geoderma* **163**, 197-208.

[13] Himes, FL, 1998 *Nitrogen, sulphur and phosphorus and the sequestration of carbon.* In R Lal et al. (eds) *Soil Processes and the Carbon Cycle.* CRC Press, Boca Raton, FL. Pp315 – 319.

[14] Williams, CH and Donald, CM 1957 Changes in organic matter and pH in a podzolic soil as influenced by subterranean clover and superphosphate. *Aust J Ag Res* **8**, 179-89.

[15] Shen, J, Yuan, L, Zhang, J, Li, H, Bai, Z, Chen, X, Zhang, W and Zhang F. 2011 Phosphorus dynamics: from Soil to Plant *Plant Physiol* **156** 997–1005.

[16] Moody, PW and Bolland, MDA 1999 *Phosphorus.* In Peverill, KI, Sparrow, LA and Reuter, DJ (eds). *Soil Analysis - an interpretation manual.* CSIRO Publishing Collingwood Australia.

[17] McLaughlin, MJ, McBeath, TM Smernik, R, Stacey, SP, Ajiboye, B and Guppy, C 2011 The chemical nature of P accumulation in agricultural soils – implications for fertiliser management and design: an Australian perspective. *Plant Soil* **349**, 69 – 87.

[18] Watt, M., Kirkegaard, JA, Passioura, JB 2006 Rhizosphere biology and crop productivity—a review Aust *J Soil Res* **44**, 299 – 317.

[19] Murphy, DV, Cookson, WR, Brainbridge, M, Marschner, P, Jones, DL, Stockdale, EA and Abbott, LK 2011 Relationships between soil organic matter and the soil microbial biomass (size, functional diversity and community structure) in crop and pasture systems in a semi-arid environment). *Soil Res* **49**, 582 – 94.

[20] Krull, ES, Skjemstad, JO and Baldock, JA. 2004 Functions of Soil organic matter and the effect on soil properties. CSIRO Land and Water, PMB2, Glen Osmond SA 5064. GRDC Project No CSO 00029. Residue, Soil Organic Carbon and Crop Performance.

[21] Hallsworth, EG and Wilkinson, GK 1958 The contribution of clay and organic matter to the cation exchange capacity of soils *J Agr Sci* **51** 1 – 3.

[22] Chan, KY, Roberts, WP, Heenan, DP 1992 Organic carbon and associated soil properties of a red earth after 10 years of rotation under different stubble and tillage practices. *Aust J Soil Res* **30**, 71 – 83.

]23] Aitken, RL, Moody, PW and McKinley PG 1990 Lime requirement of acidic Queensland soils. I relationships between soil properties and pH buffer capacity. *Aust J Soil Res* **28**, 695

[24] Helyar, KR, Cregan, PD and Godyn, DL 1990 Soil acidity in New South Wales – current pH values and estimates of acidification rates. *Aust J Soil Res* **28**, 523 – 37.

[25] Fenton, G and Helyar, K. 2007 Soil acidification. In Peter EV Charman and Brian W Murphy (eds). *Soils-Their Properties and Management.* 3rd edition. Oxford University Press, Melbourne.

IOP Conf. Series: Earth and Environmental Science **25** (2015) 012009 doi:10.1088/1755-1315/25/1/012009

Soil development over millennial timescales – a comparison of soil chronosequences of different climates and lithologies

D Sauer[1], I Schülli-Maurer[2], S Wagner[3], F Scarciglia[4], R Sperstad[5], S Svendgård-Stokke[5], R Sørensen[6], G Schellmann[7]

[1] Institute of Geography, Dresden University of Technology, Dresden, Germany

[2] Institute of Soil Science and Land Evaluation, Hohenheim University, Stuttgart, Germany

[3] Institute of Crop Science and Resource Conservation, Soil Science Division, University of Bonn, Germany

[4] Department of Biology, Ecology and Earth Sciences, University of Calabria, Arcavacata di Rende (CS), Italy

[5] The Norwegian Forest and Landscape Institute, Ås, Norway

[6] Department of Plant and Environmental Sciences, Norwegian University of Life Sciences, Ås, Norway

[7] Institute of Geography, University of Bamberg, Bamberg, Germany

E-mail: daniela.sauer@uni-hohenheim.de

Abstract. This paper reports soil development over time in different climates, on time-scales ranging from a few thousand to several hundred thousand years. Changes in soil properties over time, underlying soil-forming processes and their rates are presented. The paper is based on six soil chronosequences, i.e. sequences of soils of different age that are supposed to have developed under the similar conditions with regard to climate, vegetation and other living organisms, relief and parent material. The six soil chronosequences are from humid-temperate, Mediterranean and semi-arid climates. They are compared with regard to soil thickness increase, changes in soil pH, formation of pedogenic iron oxides (expressed as Fe_d/Fe_t ratios), clay formation, dust influx (both reflected in clay/silt ratios), and silicate weathering and leaching of base cations (expressed as (Ca+Mg+K+Na)/Al molar ratios) over time. This comparison reveals that the increase of solum thickness with time can be best described by logarithmic equations in all three types of climates. Fe_d/Fe_t ratios (proportion of pedogenic iron Fe_d compared to total iron Fe_t) reflects the transformation of iron in primary minerals into pedogenic iron. This ratio usually increases with time, except for regions, where the influx of dust (having low Fe_d/Fe_t ratios) prevails over the process of pedogenic iron oxide formation, which is the case in the Patagonian chronosequences. Dust influx has also a substantial influence on the time courses of clay/silt ratios and on element indices of silicate weathering. Using the example of a 730 ka soil chronosequence from southern Italy, the fact that soils of long chronosequences inevitably experienced major environmental changes is demonstrated, and, consequentially a modified definition of requirements for soil chronosequences is suggested. Moreover, pedogenic thresholds, feedback systems and progressive versus regressive processes identified in the soil chronosequences are discussed.

1. Introduction

Assessment of soil-forming processes and their rates is an essential scientific base for several reasons. First, it reveals the time scales on which soil-forming processes operate and thus also the time required for soil regeneration after disturbance, erosion or degeneration. Soils are a main component of terrestrial ecosystems and the base of human and animal nutrition. Their extension on the earth's surface is limited, as the earth's surface area itself is limited, and because the major part of the earth's surface is covered by water or ice. If only soils are considered that are suitable for crop production, the area is further reduced by deserts and high mountain regions. The remaining soil area can hardly be increased by any measures (with few exceptions in the Netherlands and Australia's Sunshine Coast, where extreme efforts are undertaken to gain land from the sea). In fact, the area of fertile soils steadily decreases in support of housing, logistical and industrial constructions, and due to inadequate land-use or over-use, resulting in soil degradation and erosion. At the same time, the world's population continues to increase. Since soils are an essential and limited resource it is important to understand how they form, which processes are involved in their formation, at which rates these processes proceed and lead to measurable changes in soil properties, and finally, which factors may influence the direction and rates of soil-forming processes.

Second, the quantitative assessment of the direction and rates of soil-forming processes in different climates is an important tool for the correct paleo-environmental interpretation of paleosols. The environmental conditions and the time-span in which a paleosol developed, can only be retraced, if the reconstruction is based on profound actualistic studies on soil development over time under well-defined climatic and environmental conditions. Paleosols in turn are archives, in which responses of former soils to former climatic changes are documented. Therefore, their analysis and interpretation may help us understand causal relationships between climatic shifts and the responses of soils.

Third, quantitative studies of soil development over time provide a scientific base for modelling soil formation. Modelling soil formation, in turn, is an essential requirement to take the step from description and analysis of soil development towards prediction of future soil development. Such prediction is particularly important with regard to possible soil responses to the climatic shifts that are going on at present and expected in the near future. Soil responses to these shifts will probably be inconsiderable in the central regions of ecotones, such as in most parts of the humid-temperate climate of central Europe. In contrast, serious soil responses are to be expected in the peripheral regions of more fragile ecotones, such as in dry regions of the Mediterranean (e.g. on the Iberian Peninsula), which are supposed to become warmer and drier in the coming decades and that are already at the edge of desertification. Modelling soil responses to climatic changes may identify areas at particular risk of soil degradation; hence measures (e.g. extensification and adaption of land-use) may be taken in time to prevent desertification.

Since it is impossible to directly observe soil formation over thousands of years, the direction and rates of soil-forming processes are usually assessed using soil chronosequences, i.e. sequences of soils of different ages that are supposed to have developed under similar conditions with regard to climate, vegetation and other living organisms, relief, and parent material. Important research on soil chronosequences has been done in the Mediterranean climate of California, mainly from the 1970-1980's onwards, by Harden [1, 2], who examined the Merced River soil chronosequence in central California. There, Harden also developed the profile development index (PDI) that has since been widely applied. Other significant soil chronosequence studies in California include those of McFadden and Hendricks [3] who analysed pedogenic iron forms in Holocene to middle Pleistocene soils on fluvial deposits; Busacca [4] who studied soils ranging from 600 yr to 1.6 Ma in age at Honcut Creek, Sacramento Valley; Muhs [5] who investigated a soil chronosequence on Quaternary marine terraces (< 3 ka to > 1 Ma), on San Clemente Island; and Aniku and Singer [6] who analysed pedogenic iron in a soil chronosequence on marine terraces (105-600 ka). Soil chronosequences in semi-arid climates were extensively studied in the southern Great Basin (USA), especially at Silver Lake Playa, Cima Volcanic Field, Kyle Canyon, and Fortymile Wash, by McFadden et al. [7, 8]; Reheis et al. [9]; and Harden et al. [10].

In the Mediterranean region of Europe, soil chronosequences were investigated mainly in southern Spain and Italy. Important studies were done by Alonso et al. [11] and Dorronsoro and Alonso [12] on soils on Holocene to Pleistocene fluvial terraces in Spain, and by Scarciglia et al. [13] on soils on Quaternary marine terraces in Calabria (southern Italy).

Soil chronosequences in temperate- and cool-humid climates were studied both in North America and Europe. Studies in North America include those of Singleton and Lavkulich [14] who investigated soils on beach sand on Vancouver Island; Barrett and Schaetzl [15] and Barrett [16] who examined podzolisation in sandy terraces and beach ridges at Lake Michigan; Alexander and Burt [17], and Burt and Alexander [18] who analysed soils on moraines of Mendenhall Glacier, SE Alaska. European soil chronosequence studies comprise those of Arduino et al. [19] who analysed iron oxides and clay minerals in soils on fluvial terraces in northern Italy; Bain et al. [20] who investigated soils on river terraces ranging in age from 80-13,000 years in Scotland; and Mokma et al. [21] who studied a Holocene Podzol chronosequence in Finland. Sequences of moraine ridges in mountain regions, e.g. in the Sierra Nevada (California) and in the European Alps, have also been used for studying soil chronosequences.Birkeland and Burke [22] investigated soil catena chronosequences on moraine ridges in the eastern Sierra Nevada, California, and Egli et al. [23, 24, 25] who analysed Holocene soil chronosequences in the Swiss Alps.

In contrast to the above-mentioned Mediterranean and temperate to cool climates, where numerous soil chronosequences have been studied over the last four decades, soil chronosequence studies in tropical regions and some extreme environments such as cold deserts are rare. The few existing studies include those of Pillans [26] who investigated soils on basaltic lava flows with ages ranging from 10 ka to 5.59 Ma in the tropical climate of northern Queensland (Australia), and Muhs [27] who studied soils on Pleistocene reef terraces of Barbados. These soils formed in Sahara dust, volcanic ash from the Lesser Antilles island arc, and detrital carbonate from the underlying reef limestone. A soil chronosequence in a cold desert environment was analysed by Bockheim [28] who studied soils on moraines ranging from ca. 6 ka to 250 ka in age in the Transantarctic Mountains.

This paper compares the direction and rates of soil-forming processes in humid-temperate, Mediterranean and semi-arid climates. It aims at contributing to the knowledge on natural soil-forming processes that is required for the three main purposes mentioned in the beginning of the introduction. For this purpose, six soil chronosequences are presented. In addition to the comparison of three climatically different regions, also two diverse pathways of soil formation in different parent materials in the same region in humid-temperate climate are included (Figure 1). The sites for these soil chronosequences were selected trying to avoid considerable human influence. Man has become a major soil-forming factor [29]. This factor is excluded here as far as possible, in order to ensure a good understanding of (semi-) natural soil formation first, as a base for unambiguous identification of the effect of the soil-forming factor man, separately. However, a certain extent of human influence could not be avoided, especially in the Mediterranean soil chronosequence where the soils are largely influenced by long-term agricultural use and related erosion.

Similarities and differences between the soil chronosequences in terms of identified soil-forming processes and their rates are discussed. Using the example of the Mediterranean soil chronosequence from southern Italy, comprising ca. 730 ka, the fact that soils of long chronosequences inevitably experienced major environmental changes is demonstrated, and, consequentially a modified definition of requirements for soil chronosequences is suggested.

2. Material and methods

Three soil chronosequences were studied in southern Norway. Two of them were on marine loamy sediments in the provinces Vestfold and Østfold, on both sides of the Oslo Fjord, and one was on beach sand in Vestfold (Tables 1-3). These areas have been subject to continuous glacio-isostatic upward movement, since the ice sheet in the southern Oslo Fjord region melted between ~13000 and 12000 calendar years ago. As a consequence of the steady uplift since the weight of the ice had gone, more and more parts of formerly submerged areas have risen above sea-level. Thus, the age of the land surface continuously increases from the coast towards the inland. Since the two chronosequences on loamy sediments in Vestfold and Østfold are very similar with respect to all soil properties discussed here, they have been merged into one sequence comprising 12 pedons in this paper. More details on soil properties and analytical data of the Norwegian soil chronosequences were previously reported [30, 31, 32, 33, 34]. A Mediterranean soil chronosequence was studied at the Gulf of Taranto near Metaponto, on the Ionian coast of Basilicata, southern Italy (Table 4). This area is subject to tectonic uplift due to the collision of the African and Eurasian plates. The interplay between land uplift and sea level fluctuations due to alternating glacial and interglacial periods during the Pleistocene led to a sequence of 11 marine terraces [35]. Details on the age estimates of the terraces and on the soils were previously published [36]. Two soil chronosequences in semi-arid climate were located on Holocene beach ridges along the eastern coast of Patagonia, in southernmost South America, specifically in the northern part of Golfo San Jorge. Soil data of these sequences were reported in [37].

This paper thus compares soil chronosequences that were individually described in a number of publications with regard to observed changes in soil properties over time, including soil thickness increase, changes in soil pH, formation of pedogenic iron oxides (expressed as Fe_d/Fe_t ratios), clay formation, dust influx (both reflected in clay/silt ratios), and silicate weathering and base leaching (expressed as (Ca+Mg+K+Na)/Al molar ratios) over time.

Figure 1. Soil development a) in the humid-temperate climate of S Norway (left: Podzol development on beach sand, right: Albeluvisol development on loamy marine sediments), b) in the Mediterranean climate of southern Italy and c) in the semi-arid climate of eastern Patagonia.

IOP Conf. Series: Earth and Environmental Science **25** (2015) 012009 doi:10.1088/1755-1315/25/1/012009

All soils were described according to FAO [38] and classified according to WRB [39] (Tables 1-6).

Table 1. Soil chronosequence on beach sand in Vestfold, southern Norway (MAT 6°C; MAP 975 mm)

Pedon (no., name, location)	Age[1]	Horizon sequence (depth in cm, horizons according to FAO [38]	Classification according to WRB [39]
Pedon 1 SVF2.4 (Gleåbukta)	2 300 ± 150	organic layer: 0-+5 Oi; 0-8 Ah1; 8-25 Ah2; 25-46 Bw; 46-56 BC; 56-115 Cg1; 115-135 2Cg2	Endoskeletic Brunic Endostagnic Umbrisol (Arenic, Colluvic)
Pedon 2 SVF3.0 (Gleåbukta)	3 800 ± 150	organic layer: 0-+3 Oa; +3-+8 Oe; +8-+9 Oi; 0-7 AE; 7-16 Bw1; 16-21 Bw2; 21-37 BC; 37-47 Ahb; 47-77 BC; 77-124 BCg; 124-138 2Cg; 138-180 Cr	Brunic Arenosol (Colluvic, Protospodic)
Pedon 3 SVF4.0 (Haraldsrød)	4 300 ± 200	organic layer: 0-+0.5 Oe; +0.5-+3.5: Oi 0-4 Ah; 4-9 AE; 9-22 Bh; 22-36 Bs; 36-50 Bw; 50-90 BCg; >90 2R	Endostagnic Endoleptic Cambisol (Protospodic)
Pedon 4 SVF6.0 (Jåberg)	6 600 ± 170	organic layer: 0-+3.5 Oa; 3.5-+9 Oe; +9-+13 Oi; (0-5 AE)[2]; 0/5-13 EA; 13-26 Bhs; 26-54 Bs; 54-74 BC; 74-85 Cg	Endostagnic Folic Podzol
Pedon 5 SVF7.2 (Grønneberg)	7 650 ± 130	organic layer: 0-+2 Oa; +2-+6.5 Oe; +6.5-+7.5 Oi; 0-10 AE; 10-24 Bsh; 24-40 Bs1; 40-65 Bs2; 65-105 BCg; 105-140 2Cg	Endostagnic Podzol
Pedon 6 SVF8.5 (Rauan)	9 650 ± 100	organic layer: 0-+3.5 Oa; +3.5-+8.5 Oe; +8.5-+10 Oi 0-5 AE; 5-7 E; 7-30 Bs; 30-76 2Bs2; 76-105 3Cg	Endoskeletic Endostagnic Folic Podzol (Ruptic)

[1] ages derived from calibrated ^{14}C datings, given in calendar years before sampling, not BP (i.e. before 1950)
[2] discontinuous horizon

Table 2. Soil chronosequence on loamy marine sediments in Vestfold, southern Norway (MAT 5.3-6.6 °C; MAP 880-1075 mm)

Pedon (no., name, location)	Age[1]	Horizon sequence (depth in cm, horizons according to FAO [38]	Classification according to WRB [39]
Pedon 1 VF2.4 (Sem)	1 650 ± 120	0-5 Ah; 5-13 BE; 13-20 E1; 20-40 E2; 40-56 Btg1; 56-95 Btg2; 95-130 Ctr; 130-155 Cr	Luvic Endogleyic Stagnosol (Siltic)
Pedon 2 VF4.5 (Ramnes)	4 600 ± 70	0-9 Ah; 9-21 Ap; 21-27 E; 27-40 Beg; 40-70 Btg1; 70-100 Btg2; 100-135 Btg3; 135-170 Cg; 170-230 C	Luvic Stagnosol (Siltic)
Pedon 3 VF6.6 (Fossan)	6 200 ± 100	0-9 Ah; 9-15 Ap; 15-30 BE; 30-42 E; 42-56 E/Bg; 56-91 Btg1; 91-195 Btg2; 195-230 C	Luvic Glossic Albic Fragic Stagnosol (Siltic)
Pedon 4 VF8.8 (Holmen)	6 550 ± 200	0-8 Ah; 8-20 Ap; 20-36 BE; 36-40 E; 40-66 E/Bg; 66-90 B/Eg; 90-113 Btg1, 113-155 2Btg2; 155-175 3Cg	Luvic Glossic Albic Fragic Stagnosol (Siltic)
Pedon 5 VF7.3 (Gjein)	8 100 ± 120	0-4 Ah; 4-18 BE; 8-27 E; 27-40 E/Bg; 40-58 Btg1; 58-105 Btg2; 105-190 Cg	Luvic Glossic Albic Fragic Stagnosol (Endoeutric)
Pedon 6 VF9 (Torp)	9 000 ± 150	0-6 AE; 6-9 Bs; 9-24 BE; 24-27 E; 27-49 E/Bg; 49-85 Btg; 85-156 Bg; 214 Cg	Luvic Glossic Albic Fragic Stagnosol (Protospodic)

[1] ages derived from calibrated ^{14}C datings, given in calendar years before sampling, not BP (i.e. before 1950)

Table 3. Soil chronosequence on loamy marine sediments in Østfold, southern Norway (MAT 4.6-6.4 °C; MAP 770-880 mm)

Pedon (no., name, location)	Age[1]	Horizon sequence (depth in cm, horizons according to FAO [38]	Classification according to WRB [39]
Pedon 1 ØF3 (Løkkevika)	3 000 ± 250	0-8 Ah; 8-30 BE; 30-40 E; 40-80 Btg; 80-110 Bg; 110-155 Cr1; 155-210 Cr2	Hyperdystric Alic Endogleyic Stagnosol (Siltic)
Pedon 2 ØF4 (Tomb)	3 500 ± 200	0-8 Ah; 8-25 Ap; 25-40 E; 40-77 Btg; 77-116 Btrg; 116-150 Cr	Luvic Albic Endogleyic Stagnosol (Endosiltic)
Pedon 3 ØF7.5 (Husevja)	6 550 ± 150	0-3 Ah; 3-12 AE; 12-26 Bs; 26-29 BE; 29-38 E; 38-66 E/Bg; 66-80 Btg; 80-105 Cg	Luvic Glossic Albic Fragic Stagnosol (Episiltic, Protospodic)
Pedon 4 ØF5 (Navestad)	6 650 ± 150	0-6 Ah; 6-16 Ap; 16-25 E; 25-40 E/Bg; 40-66 Btg1; 66-90 Btg2; 90-170 Cr1; 170-185 Cr2	Luvic Glossic Albic Endogleyic Fragic Stagnosol (Endoeutric)
Pedon 5 ØF8 (Os Kirke)	9 750 ± 150	0-5 Ah; 5-13 EB; 13-30 E1; 30-40 E2; 40-70 E/Bg; 70-127 Btg; 127-175 BCg; 175-195 Cg	Luvic Glossic Albic Fragic Stagnosol (Endoeutric, Siltic)
Pedon 6 ØF11 (Båstad)	11 050 ± 150	0-6 Ah; 6-10 AE; 10-21 Bs; 21-37 E; 37-60 E/Bg; 60-77 Btg; 77-90 BCg1; 90-100 BCg2; 100-175 BCg3; 175-195 C	Luvic Glossic Albic Fragic Stagnosol (Endofluvic, Siltic, Protospodic)

[1] ages derived from calibrated ^{14}C datings, given in calendar years before sampling, not BP (i.e. before 1950)

Table 4. Soil chronosequence on marine terraces near Metaponto, Gulf of Taranto, southern Italy (MAT 16.2 °C; MAP 456 mm at Marina di Ginosa)

Pedon (no., name, location)	Age[1]	Horizon sequence (depth in cm, horizons according to FAO [38]	Classification according to WRB [39]
Pedon 1 T0; Lido di Metaponto	190[1]	0-2 C; 2-13 AC; 13-31 CB1; 31-46 2CB2; 46-67 Cw1; 67-95 Cw2; 95-158 C	Calcaric Brunic Arenosol (Ochric)
Pedon 2 T1; Petrulla (Loess)	16 000[1]	0-30 Ap; 30-60 Bw; 60-120 Bk1; 120-185 Bk2; 185-240 2Bk3; 240-450 3C	Endocalcic Luvisol (Loamic)
Pedon 3 T2; San Teodoro I	100 000	0-23 Ap; 23-60 Bt1; 60-93 Bt2; 93-130 Ck1; 130-180 Ck2; 180-280 2Bk1; 280-325 3Bk2; 325-350 4Bk3; 350-400 Ckm; >400 5Cw	Endocalcic Luvisol (Cutanic)
Pedon 4 T3; San Teodoro II	120 000[2]	0-25 Ap; 25-70 Ap/Bt (tilted); 70-100 Bt/Ap (tilted); 100-112 Bt; >112 Bk	Bathicalcic Luvisol (Cutanic)
Pedon 5 T4; Marconia	195 000	0-25 Ap1; 25-42 Ap2; 42-58 Bt1; 58-116 2Btk1; 116-133 Btk2; 133-151 Bt2; 151->223 3Bt3	Endocalcic Alisol (Clayic, Cutanic)
Pedon 6 T5-SE	310 000	0-30 Ap; 30-54 Bt1; 54-83 Bt2; 83-108 Btk1; 108-174 2Bt3; 174->206 Btk2	Endocalcic Alisol (Cutanic)
Pedon 7 T5-NW	330 000	0-33 Ap; 33-80 Bt; 80-127 2Btk1; 127-168 Btk2; 168->205 Btk3	Endocalcic Alisol (Cutanic)
Pedon 8 T6; Tinchi I	405 000	0-20 Ap; 20-36 Bt1; 36-50 2Bt2; 50-108 Bt3; 108-165 Bw1; 165-182 3Bw2; 182-215 Cw; 215->250 C	Chromic Luvisol (Clayic, Cutanic)
Pedon 9 T7; Tinchi II	500 000	0-33 Ap; 33-54 Bt1; 54-73 Bt2; 73-160 2Bt3; 160-290 Bt4; 290-330 Bw; 330-338 Bg; 338-380 Ck/Ckm; 380-410 3C	Chromic Luvisol (Cutanic)
Pedon 10 T8; Pisticci	575 000[3]	0-22 Ap; 22-47 AB; 47-65 2Btk1; 65-109 Btk2; 109-200 Btk3; 200-239 Btk4; 239-259 Btk5; 259-286 CB1; 286-323 CBk; 323->380 CB2	Chromic Alisol (Cutanic)
Pedon 11 T9; Bernalda	670 000	0-10 Ah; 10-26 Ah/Bw; 26-70 Bt; 70-100 Btk1; 100-117 Btk2; 117-140 Btk3; 140-220 Btk4; 220->230 2Bw	Chromic Luvisol (Cutanic)

[1] OSL dating from [40]; [2] age based on occurrence of Senegalese fauna (Eemian); age based on tephra chronology; all other ages estimated according to Pleistocene glacial-interglacial cycles and related sea level fluctuations

Table 5. Soil chronosequence on beach ridges with substantial dust accumulation along the Patagonian East coast (MAT 12.6 °C; MAP 287 mm)

Pedon (no., name, location)	Age[1]	Horizon sequence (depth in cm, horizons according to FAO[38]	Classification according to WRB [39]
Pedon 1 R1 (Bust.); Pa'04/12b	2241 ± 46	0-1 (dp); 1-4 Ah; 4-8 AC1; 8-18 AC2; 18-30 C(h); 30-42 C(h)k; 42-50 2Ck1; 50-60 2Ck2; 60-73 2Ck3; 73-100 3C	Calcaric Skeletic Regosol (Protocalcic, Yermic)
Pedon 2 R2 (Cam.); Pa'04/8	3372 ± 48	0-1 (dp); 1-4 Ah; 4-26 AC; 26-55 C(h)k; 55-65 2Ck; 65-75 2C	Calcaric Skeletic Regosol (Protocalcic, Yermic)
Pedon 3 R3 (Bust.); Pa'04/4e	4052 ± 50	0-1 (dp); 1-4 Ah1; 4-17 Ah2; 17-44 AC; 44-65 C(h)k1; 65-80 2C(h)k2; 80-87 2Ck1; 87-100 2Ck2; 100-140 2C	Calcaric Skeletic Regosol (Protocalcic, Yermic)
Pedon 4 R4 (Bust.); Pa'04/2	6238 ± 51	0-1 (dp); 1-5 Ah; 5-20 AC; 20-60 C(h); 60-95 2C1; 95-100 2Ck1; 100-135 2C2; 135-139 2Ck2; 139-160 2C3; 160-164 2Ck3; 164-176 2C4	Calcaric Skeletic Regosol (Protocalcic, Yermic)

[1] ages are uncalibrated radiocarbon ages of molluscs from each beach ridge

Table 6. Soil chronosequence on beach ridges with minor dust accumulation along the Patagonian East coast (MAT 12.6 °C; MAP 287 mm)

Pedon (no., name, location)	Age[1]	Horizon sequence (depth in cm, horizons according to FAO [38]	Classification according to WRB [39]
Pedon 1 L1 (Cam.); Pa'04/6	1376 ± 47	0-3 (dp)[3]; 3-5 Ah; 5-8 AC; 8-22 C(h); 22-50 2C	Calcaric Hyperskeletic Leptosol (Yermic)
Pedon 2 L2 (Cam.); Pa32	2618 ± 92	0-1 (dp); 1-4 Ah; 4-25 AC; 25-30 2Ck; 30-60 2C	Calcaric Hyperskeletic Leptosol (Protocalcic, Yermic)
Pedon 3 L3 (Bust.); Pa'04/1b	5400 ± 25	0-2 (dp); 2-7 Ah; 7-20 C(h)1; 20-50 C(h)2; 50-80 Ck; 80-100 C	Calcaric Hyperskeletic Leptosol (Protocalcic, Yermic)

[1] ages are uncalibrated radiocarbon ages of molluscs from each beach ridge

2. Comparison of soil property changes in the different chronosequences

2.1. Soil thickness increase

Soil thickness was calculated according to [12] as thickness of A horizons + thickness of B horizons + ½ thickness of transitional AC/CA/BC/CB horizons. Soil thickness increase over time can be best described by logarithmic functions in all studied soil chronosequences (Figure 2). The rates of increase in soil thickness are clearly influenced by climate and parent material. The highest rate is observed for the Mediterranean climate (Figure 2; green triangles). The humid-temperate climate of southern Norway allows for medium rates of soil thickness increase over time, whereby rates in the loamy marine sediments (Figure 2; red squares) are higher than those in the sandy beach deposits (Figure 2; orange squares). The lowest rates of soil thickness increase are found in the semi-arid climate of Patagonia (Figure 2; green diamonds), where the mean annual temperature (12.6 °C) is in between those of the two other regions, but the low amount of precipitation (MAP = 287 mm) limits the rates of weathering and primary plant and animal productivity, and increase in soil thickness.

In general, it was attempted to establish all soil profiles comprised in the chronosequences in flat positions, showing minimum erosion. This approach worked well in the Holocene chronosequences in Norway and Patagonia, whereas in the Holocene-Pleistocene chronosequence in southern Italy, spanning ~730 ka, it was impossible to include only soils that were not or only slightly eroded because erosion was ever-present. A theoretical function of increase in soil thickness over time was obtained by including only maximum soil thicknesses (Figure 3). However, very few of the soils included in this curve were really not eroded, hence without erosion, still higher rates of soil thickness increase would be observed.

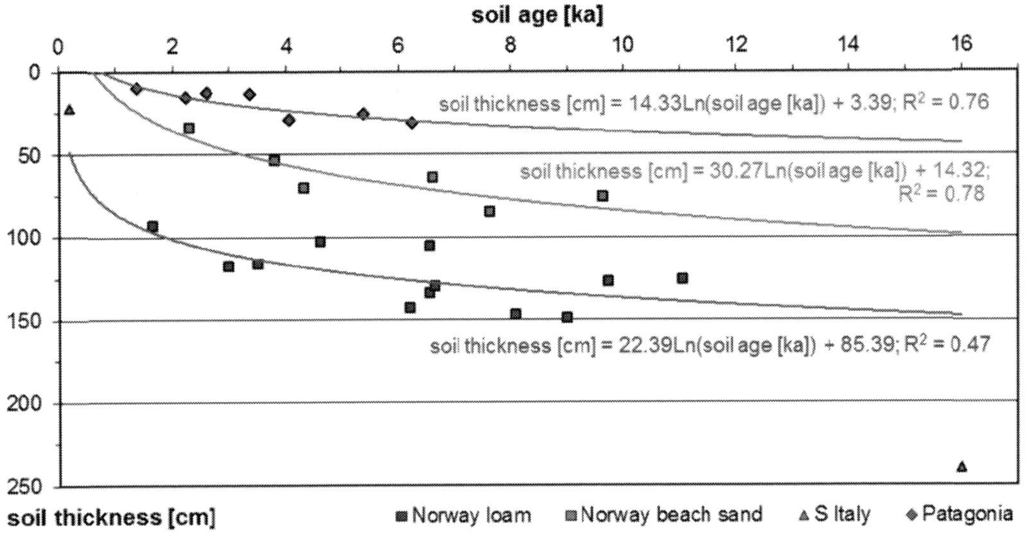

Figure 2. Soil thickness increase over a time-span of 16 ka in the studied soil chronosequences. The two soil chronosequences on loamy sediments in Norway and on gravelly beach ridges in Patagonia could be merged with regard to soil thickness.

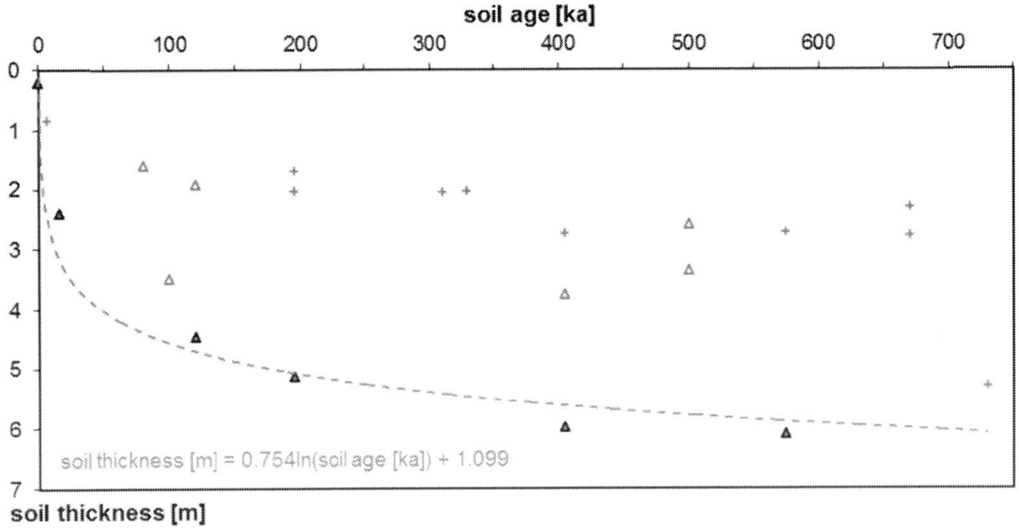

Figure 3. Soil thickness increase observed in the Italian soil chronosequence over a time-span of 730 ka. Open triangles = eroded soils; + = C horizon not reached.

2.2. Soil pH

Weighted mean soil pH values of the upper 50 cm of the soils were calculated by taking the anti-log of the pH value for each horizon down to 50 cm depth and multiplying the resulting proton concentration by horizon thickness; the products were summed up and divided by 50.

The different soil moisture regimes of the study areas are clearly reflected in the pH levels (Figure 4). The study areas in the humid-temperate climate in Norway are characterised by distinct leaching environments. Although pedogenesis follows different pathways in the loamy marine sediments and sandy beach deposits, the pH values are very similar in the Norwegian soil chronosequences (Figure 4; orange and red squares). Within a very short period of time carbonates, which may be present in the sediments in the beginning (mainly in the form of shell fragments), are dissolved. This process is favoured by the presence of sulphides in the loamy sediments, and by the high water conductivity in the sandy sediments, respectively. After complete removal of carbonates, pH drops rapidly to around pH 4 and stays rather constant thereafter, due to buffering by adsorption of protons to clay minerals and proton consumption in weathering and clay mineral formation. Due to these buffering systems, after ~2 ka, pH in the Norwegian soils is influenced by the local vegetation rather than soil age.

The Mediterranean climate, in which the Italian soil chronosequence is located, corresponds to a xeric soil moisture regime. This means that leaching conditions prevail during the moist winter months, whereas there are no leaching conditions in the dry summers. Such an environment should, over a long time-span, gradually lead to decarbonatisation (removal of carbonates) and pH drop but at distinctly lower rates than in a humid climate. However, there is another important factor counteracting decarbonatisation; the deposition of considerable amounts of dust characteristic in the Mediterranean region, which leads to accumulation of carbonates and re-carbonatisation of the soils. The rates of dust influx must have varied through time. Well-developed Bt horizons with thick illuvial clay films, which are typical for Mediterranean soils on calcareous parent materials (including those comprised in the Italian chronosequence), indicate that there must have been periods with low dust influx, allowing for complete decarbonatisation and clay illuviation. On the other hand, Mediterranean soils often show accumulation of secondary carbonates within previously decarbonated horizons, indicating phases of increased influx of calcareous dust (Figure 5).

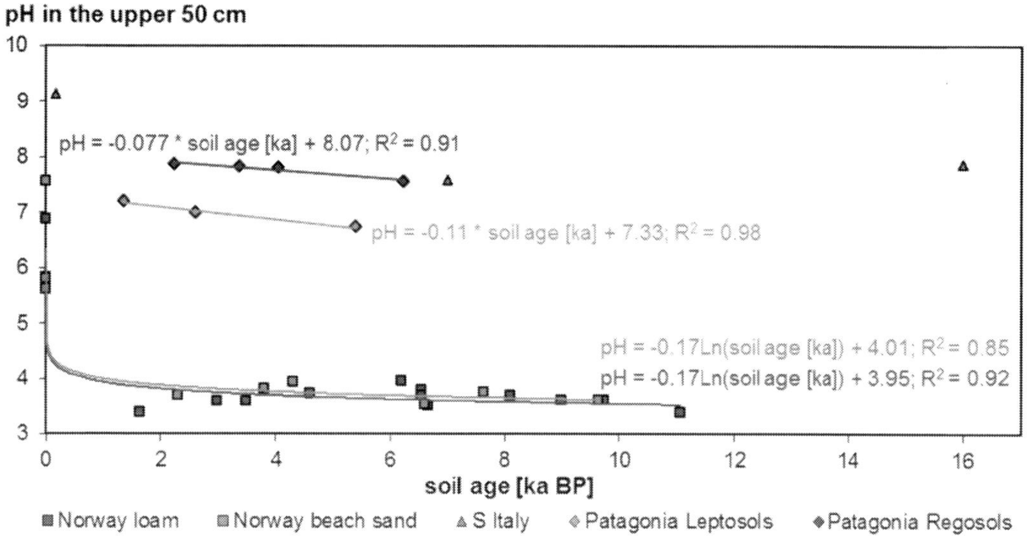

Figure 4. Soil pH, weighted means of the upper 50 cm (soil:solution ratio = 1:2.5; measured in KCl for Norway and Italy and in $CaCl_2$ for Patagonia). The two soil chronosequences on loamy sediments in Norway could be merged with regard to soil pH.

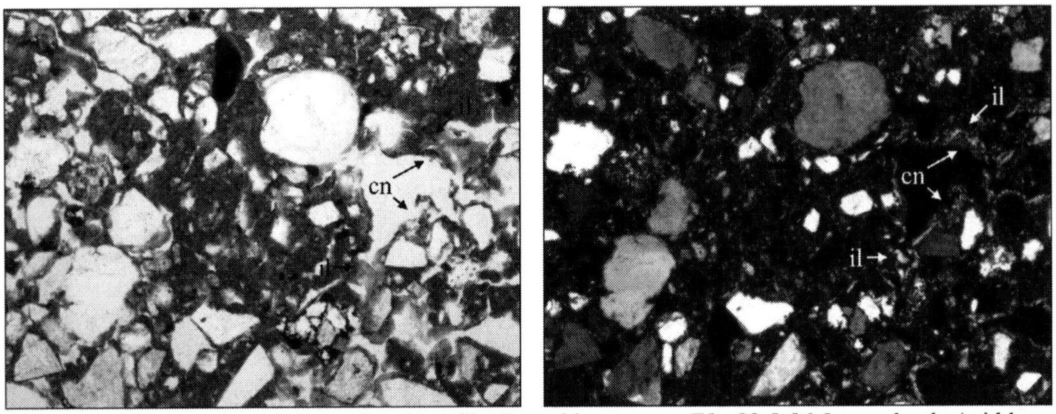

Figure 5. Thin section photograph of profile Bernalda, terrace T9, 82.5-86.5 cm depth (width = 2.2 mm; left with plane polarised light, right with crossed polarisers). The middle Pleistocene soils of the Italian chronosequence are characterised by neo-formed reddish brown clay and illuvial clay (il). This soil (ca. 670 ka-old) moreover shows accumulation of secondary calcite needles (cn) in many voids.

Because of the xeric moisture regime and variable re-carbonatisation through dust influx, the general pH trend with time in the Metaponto chronosequence shows a slow decrease with some scatter and even increased pH values in the older soils of the sequence (Figure 6). Pleistocene alternations between Mediterranean forest (during interglacial periods) and steppe environments (during glacial periods) in southern Italy [41, 42] must have been accompanied by alternating leaching and non-leaching or slight leaching conditions. However, the influence of paleo-environmental changes on the time courses of soil chemical parameters is difficult to reconstruct. It is likely that the rates of decarbonatisation and pH decrease in the upper 50 cm were considerably decreased and even temporarily reversed during glacial periods with steppe environments and enhanced dust influx rates. Also the environmental conditions within glacial periods were not constant but included, for example shifts between *Artemisia* steppe and forest steppe, which must have been related to different soil moisture regimes and hence different carbonate dynamics.

The Patagonian soil chronosequences are located in the driest environment included in this work. However the gravelly beach ridges, on which the soil chronosequences in Patagonia were established, are highly water permeable. Therefore, decarbonatisation is taking place in the upper 20 to 40 cm of these soils, and pH decreases towards the soil surface, indicating that some leaching processes are active. Both chronosequences show a slow but distinct pH decrease with time, whereby a clear difference in pH levels is observed between the soil chronosequence on beach ridges with > 10 % fine earth, i.e. the Regosol sequence (Figure 4; dark blue diamonds), and the sequence on beach ridges with ≤ 10 % fine earth in the upper 75 cm, i.e. the Leptosol sequence (Figure 4; light blue diamonds). The pH values of the Regosols are 0.8 to 0.9 pH units higher than those of the Leptosols. The higher amounts of fine earth in the Regosols point to greater dust accumulation at these sites, compared to the Leptosol sites; it can thus be concluded that the higher pH level of the Regosol sequence is also due to the higher input of calcareous dust.

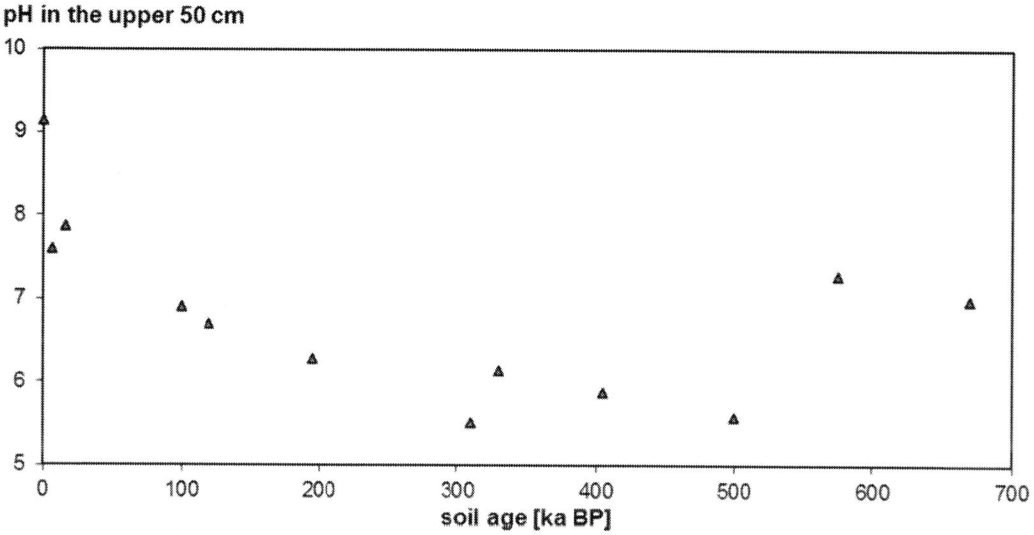

Figure 6. Soil pH in KCl (weighted means of the upper 50 cm) over the whole time-span of the Italian soil chronosequence.

2.3. Formation of pedogenic iron oxides

The formation of pedogenic iron oxides in the different soil chronosequences is compared by calculating the weighted average Fe_d/Fe_t ratio (ratio of dithionite-extractable iron, analysed by ICP-OES and total iron, analysed by X-ray fluorescence analysis) in the upper 50 cm of the soils. For this purpose, the Fe_d/Fe_t ratio of each horizon down to 50 cm depth was multiplied by horizon thickness, the results were summed up and divided by 50.

The Norwegian soil chronosequences show increasing trends of pedogenic iron oxides over time, which can be described by linear functions (Figure 7; orange and red squares). The Italian chronosequence shows a clear increase in Fe_d/Fe_t ratios in the first ~120 ka, whereas Fe_d/Fe_t ratios in the older soils are highly variable (Figure 8). Decreased Fe_d/Fe_t values (open triangles in Figure 8) are interpreted to correspond to severely eroded soils, since erosion removes the uppermost, most weathered parts of the profile, thus exposing lower, less weathered parts of the soil having lower Fe_d/Fe_t ratios. A logarithmic curve can be fitted to the maximum Fe_d/Fe_t values of the Italian chronosequence. It is concluded that: 1) the contents of pedogenic iron oxides in the middle Pleistocene soils of the Italian chronosequence are influenced to variable degree by erosion; and 2) the time course of the progressive formation of pedogenic iron oxides can be described by a logarithmic equation. However, it has to be considered that this equation averages over higher rates of pedogenic iron oxide formation in interglacial periods and lower rates in glacial periods.

The Patagonian soil chronosequences exhibit a reverse development of pedogenic iron oxide contents; their Fe_d/Fe_t ratios decrease with soil age (Figure 7; blue diamonds). This unusual time course of Fe_d/Fe_t probably reflects proceeding accumulation of dust in the semi-arid environment of these soils. Central eastern Patagonia is characterised by strong winds and sparse vegetation cover. This combination of environmental factors leads to enhanced aeolian activity. Dust that settles down on the gravelly beach ridges, on which the soil chronosequences were established, is easily trapped in the interstitial voids between the pebbles and accumulates in the soils. Dust accumulation leads to dilution of the soils with regard to Fe_d/Fe_t ratios.

2.4. Clay/silt ratios

Clay/silt ratios can be used in soil chronosequences on homogenous parent materials to assess progressive clay formation with time. However, variations in the primary clay/silt ratio of sediment layers due to slight changes in the depositional conditions (e.g. water velocity) may limit the applicability of this approach. The soil chronosequences comprised in this paper are compared with regard to the weighted mean clay/silt ratios in the upper meter where possible (Norway, Italy), because the depth of 50 cm that has been used for pH and Fe_d/Fe_t ratios is not useful for the clay/silt ratio of soils that are characterised by clay illuviation. The clay/silt ratios of the Patagonian chronosequences, however, refer to the upper 50 cm because not all pedons of these sequences were sampled down to one meter depth.

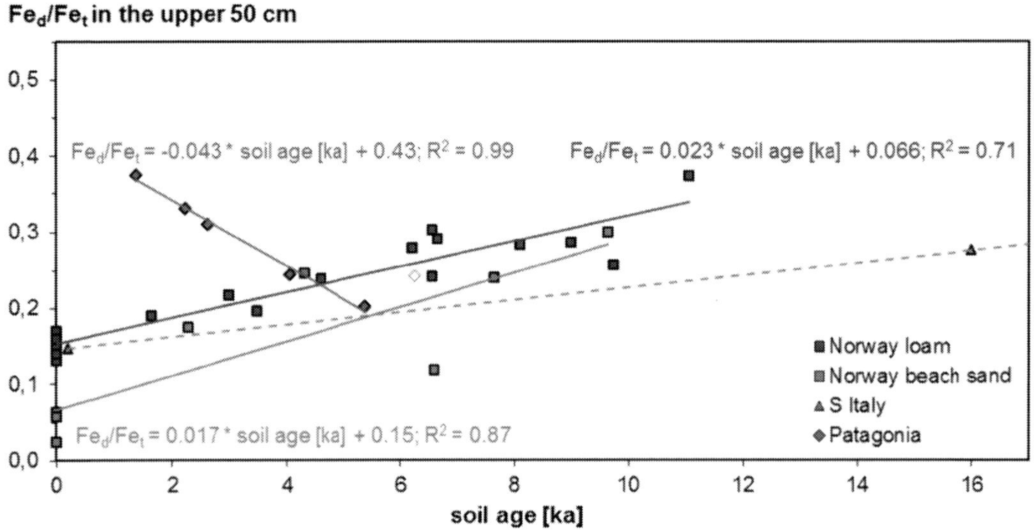

Figure 7. Trends of Fe_d/Fe_t in the studied soil chronosequences (weighted means of the upper 50 cm of each pedon). The two soil chronosequences on loamy sediments in Norway and on gravelly beach ridges in Patagonia could be merged with regard to Fe_d/Fe_t ratios.

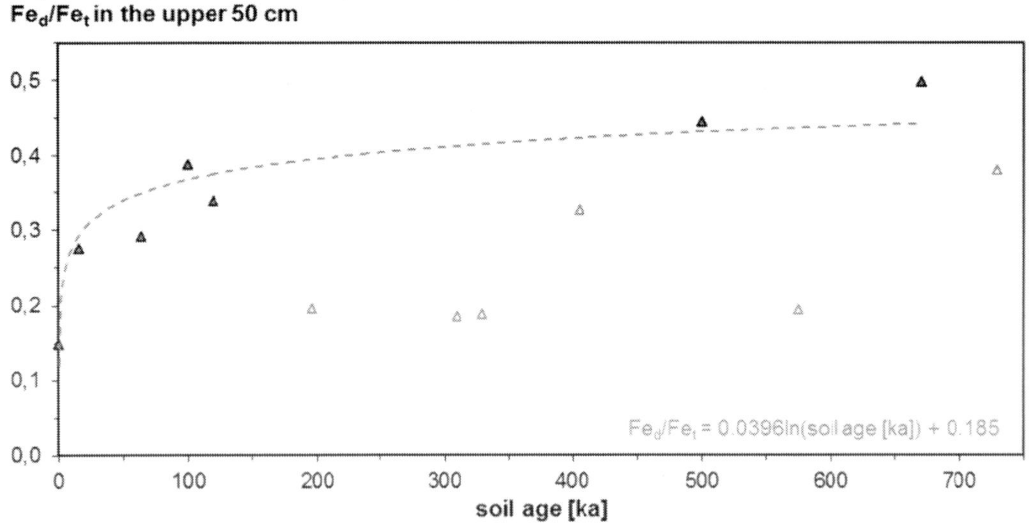

Figure 8. Fe_d/Fe_t ratios (weighted means of the upper 50 cm of each pedon) over the whole time-span of the Metaponto soil chronosequence. Open triangles represent Fe_d/Fe_t ratios of strongly eroded soils and were not included in the curve.

The soil chronosequences compared in this paper have all developed on marine, littoral and alluvial sediments. The parent materials of the pedons of the same chronosequence usually have the same source and thus similar mineralogical and geochemical composition, however, they vary with regard to texture. This variability is reflected in a large scatter of clay/silt ratios, particularly in the sequences on marine loamy sediments in Norway (Figure 9, red squares). The sandy beach deposits in Norway also show some scatter, but nevertheless a trend of increasing clay/silt ratios with time can be observed (Figure 9, orange squares). The Italian sequence exhibits a rough trend of increasing clay/silt ratios, but at the same time shows considerable scatter. The Patagonian soil chronosequences exhibit a distinct linear decrease in their clay/silt ratios (Figure 8; blue diamonds). This trend is caused by the high rate of dust influx and related silt accumulation, compared to the low rate of clay formation in the semi-arid climate of this area. This observation confirms that the proceeding dust accumulation is a main soil-forming process in this area, as has already been suggested by the decreasing Fe_d/Fe_t ratios of the Patagonian chronosequences.

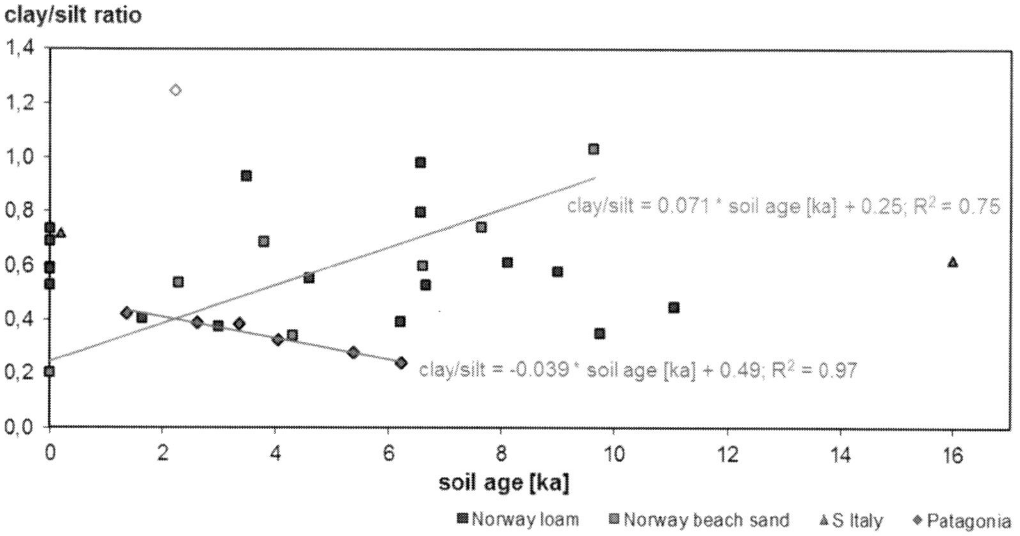

Figure 9. Clay/silt ratios, weighted mean values of the upper 100 cm (Norway and S Italy) or the upper 50 cm, respectively (Patagonia). Open blue diamond = outlier in Patagonian sequence.

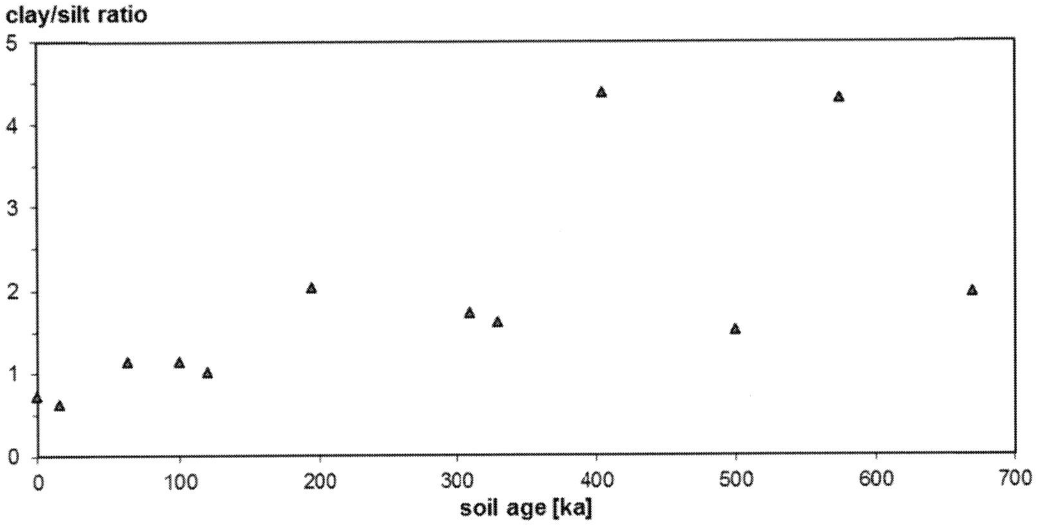

Figure 10. Clay/silt ratios (weighted mean values of the upper 100 cm) over the whole time-span of the Italian sequence.

2.5. Silicate weathering and base leaching

In general, one would expect that the main factors enhancing silicate weathering are:

1) soil moisture: moisture is required for important chemical weathering processes, e.g. for hydrolysis and for formation of acids as important weathering agents, and for leaching of weathering products from the weathering minerals from which they were released, which then allows for continuation of weathering.
2) soil temperature: temperature should increase weathering rates, since, according to the Arrhenius equation a temperature increase of 10 K induces doubling to triplication of the rates of any chemical reaction.

Several indices have been used in the literature to express the degree of silicate weathering and base leaching. Here, the molar ratio of (Ca+Mg+K+Na)/Al is used. The weighted mean values of the (Ca+Mg+K+Na)/Al molar ratios in the upper 50 cm of the soils of all chronosequences were calculated after subtracting the molar proportion of Ca from calcium carbonates.

The Norwegian soil chronosequences show large scatter of (Ca+Mg+K+Na)/Al ratios (Figure 11; orange and red squares). The pedons on beach sand generally have higher ratios than those on marine loam, because of the higher quartz and lower silicate contents of the beach sand, involving lower Al contents. Despite the scatter, it is evident that the rates of (Ca+Mg+K+Na)/Al decrease in the Norwegian sequences (Figure 11; slopes of the red and orange lines) are below the Holocene rate of (Ca+Mg+K+Na)/Al decrease of the Italian chronosequence which is located in a drier moisture regime but warmer climate (Figure 11; slope of the green dashed line). The (Ca+Mg+K+Na)/Al ratios in the Italian sequence show a clear decrease over 200 ka (Figure 12). Afterwards, decreasing rates indicate exhaustion of easily weatherable silicates. A minimal further decrease is assumed but cannot be proved due to large scatter.

The Patagonian soil chronosequences show distinct decreases in their (Ca+Mg+K+Na)/Al ratios (Figure 11; slopes of the blue lines). It is, however, unlikely that these decreases are caused by strong silicate weathering and leaching in the semi-arid environment of these chronosequences. It is more probable that this decrease is related to the proceeding dust accumulation in these soils. This assumption is confirmed by the fact that the (Ca+Mg+K+Na)/Al ratios of the Regosols, i.e. the soils with stronger dust accumulation, are overall on a lower level than those of the Leptosols. Probably, the minor amounts of sandy beach sediments in the interstitial voids between the pebbles of the beach ridges have high quartz and low silicate contents, whereas the accumulated silty dust comprises comparatively high amounts of silicates, thus strongly increasing the Al contents and decreasing the (Ca+Mg+K+Na)/Al ratios with time.

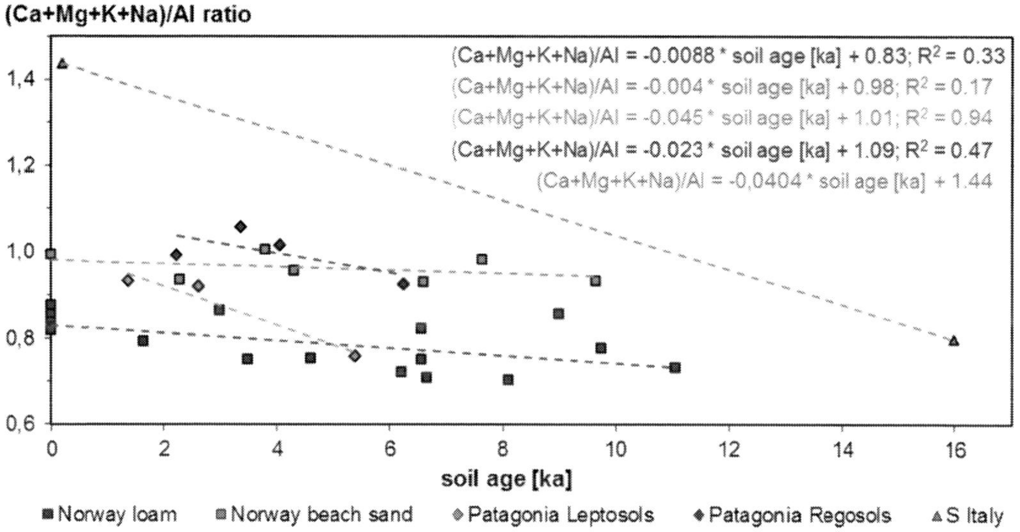

Figure 11. (Ca+Mg+K+Na)/Al molar ratio (weighted mean values of the upper 50 cm) as index for feldspar weathering and base leaching. Ca from calcium carbonate has been subtracted.

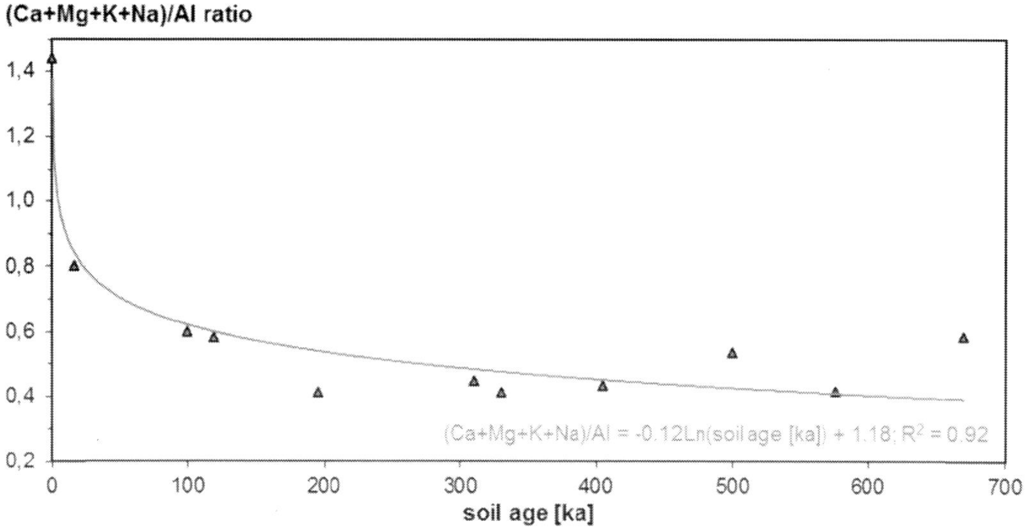

Figure 12. (Ca+Mg+K+Na)/Al molar ratio (weighted mean values of the upper 50 cm) of the Italian chronosequence over the whole time-span of the sequence.

3. Conclusions

The comparison of the six soil chronosequences shows that increase of soil thickness with time can be best described by logarithmic equations in all climates and lithologies included in this paper. Rates of soil thickness increase are influenced by both climate and parent material. Soil pH values in the upper

50 cm show decreasing trends in all climates and parent materials, whereby the general level of pH is higher in drier climates and in areas with substantial influx of calcareous dust than in humid climates. Formation of pedogenic iron oxides proceeds with time, usually leading to increasing Fe_d/Fe_t ratios, except for regions, where the influx of dust (having low Fe_d/Fe_t ratios) prevails over the process of pedogenic iron oxide formation, which is the case in the Patagonian chronosequences presented here. Dust influx has also a substantial influence on the time courses of clay/silt ratios and on element indices for silicate weathering. Variable dust influx and sediment inhomogeneity of pedons may lead to large scatter of data so that chronofunctions are difficult to obtain.

Five of the six soil chronosequences presented here are Holocene soil chronosequences, comprising time-spans of ca. 7 ka (Patagonia) to ca. 11 ka (Norway). These soils experienced some climatic shifts such as the mid-Holocene climatic optimum, and the Little Ice Age, but this climatic variation was within a range that appears to have no measureable effect on the soil chronofunctions.

In contrast, the Mediterranean soil chronosequence comprises ca. 730 ka, thus spanning a number of glacial-interglacial cycles. This means that the soil-forming factors climate and organisms were subject to major changes over the residence time of most of the soils comprised in the soil chronosequence. In the case of southern Italy, the environmental conditions alternated between Mediterranean conditions during interglacial periods and open *Artemisia* steppe to forest steppe environments during glacial periods [41, 42]. It is evident that rates of base leaching and formation of pedogenic oxides must have been significantly lower under steppe conditions than under Mediterranean conditions. It seems most likely that existing Mediterranean soils of several meters depth did not show any further deepening during periods of steppe environments. Instead, it can be assumed that enhanced dust influx and increased accumulation of soil organic matter took place in the uppermost part of the existing Mediterranean soils, and steppe soils such as Phaeozems and Chernozems formed. Each following shift to an interglacial period with Mediterranean conditions led to soil organic matter decomposition and hence degradation of the steppe soils. Mediterranean soil formation that had been interrupted during glacial periods continued, characterised by carbonate leaching, clay migration, and rubification. These conclusions are in agreement with Scarciglia et al. [13] who studied chronosequences of typical Mediterranean soils on Early to Late Pleistocene marine terraces along the northwestern coast of Calabria (southern Italy). A combination of chemical, mineralogical and micromorphological analyses revealed that soil-forming processes in those soils did apparently not proceed continuously but polycyclically. The authors thus concluded that the soils mainly developed during interglacial periods.

The problem of major changes in climate and vegetation that is discussed here, using the example of the Italian soil chronosequences, holds true for all soil chronosequences that extend back to pre-Holocene periods. Hotchkiss et al. [13], for example, discuss the impact of Pleistocene environmental changes on soil and ecosystem development in Hawaii. In such long soil chronosequences, the factor climate per se cannot be kept constant over time, as demanded in the original definition of soil chronosequences, specifying soil chronosequences as sequences of soils of different age that are supposed to have developed under similar conditions with regard to climate, vegetation and other living organisms, relief and parent material (see introduction). We thus need to accept that climate and vegetation varied over the time-spans comprised in these long soil chronosequences and that (at least) all soils of pre-Holocene age are polygenetic [29]. Hence, the requirements for these long soil chronosequences need to be modified as follows: soil chronosequences are sequences of soils of different age that are supposed to have developed under similar environmental conditions, or have experienced similar environmental changes through time, e.g. through glacial-interglacial cycles, with regard to the soil-forming factors climate, vegetation and other living organisms, relief and parent material.

In addition to the inevitable environmental variability in time discussed above, it is important to consider that, even under constant environmental conditions pedogenesis would not proceed uniformly over long time-spans. Instead, pedogenic thresholds and feedback systems naturally occur in the course of soil development. Examples for major pedogenic thresholds that occur in the soil chronosequences comprised in this paper include the: threshold of acidification and availability of organic complexing agents that has to be reached before podzolisation starts in the soil chronosequence on beach sand in Norway; and the threshold of calcium leaching that has to be reached before clay illuviation starts in the soil chronosequences on loamy sediments in Norway and in southern Italy. Feedback systems are operating in the chronosequences involving clay illuviation (in Norway and southern Italy), since clay accumulation in the developing Bt horizons will lead to a progressively finer pore system in the Bt horizons, which in turn will favour further accumulation of clay. Feedback systems are also effective in the Norwegian soil chronosequence on beach sand, where the precipitation of the first metal-organic colloids that precipitate in the subsoil will attract other colloids and thus self-enhance the development of spodic horizons. Similarly, the first carbonate crystals, precipitating in the Patagonian soils will form nuclei, on which further carbonates will precipitate.

Besides progressive pedogenic processes such as silicate weathering and formation of clay minerals and pedogenic oxides, also regressive processes need to be taken into account. The main regressive process, particularly in the Mediterranean study area, is erosion. Johnson [44] explained soil thickness as a result of the interplay of: 1) deepening through weathering; 2) upbuilding (e.g. by dust influx); and 3) erosion. According to Johnson and Watson-Stegner [45], upbuilding can be regressive if unweathered fresh material accumulates at a rate that exceeds the weathering rate, or it can be progressive if weathering keeps pace with sedimentation and the sediments do not lead to profile rejuvenation or simplification. In the case of the Italian chronosequence, upbuilding by dust deposition represents a regressive process, because dust introduces fresh carbonates to previously decarbonated and clay illuviated soils. In the Patagonian chronosequences, upbuilding by dust deposition can be regarded as a regressive process with respect to Fe_d/Fe_t and clay/silt ratios because both ratios are decreased by dust accumulation, whereas they should generally increase with proceeding soil development including neoformation of clay minerals and pedogenic iron oxides. On the other hand, the dust that accumulated in the interstitial voids between the pebbles in the Patagonian beach ridges in many sites, represents the only fine earth at all. Dust accumulation in these cases is an essential process to increase the water holding capacity of the extremely gravelly soils and allow for the succession of higher plants. From this point of view, dust accumulation in this particular environment may be rather regarded as a progressive, than a regressive, process.

References

[1] Harden JW 1982 A quantitative index of soil development from field description: examples from a chronosequence in central California. *Geoderma* **28** 1–28
[2] Harden JW 1987 Soils developed in granitic alluvium near Merced, California. *US Geol. Survey Bull.* 1590-A
[3] McFadden LD, Hendricks DM 1985 Changes in the content and composition of pedogenic iron oxyhydroxides in a chronosequence of soils in southern California. *Quat. Res.* **23** 189–204
[4] Busacca AJ 1987 Pedogenesis of a chronosequence in the Sacramento Valley, California, U.S.A., I. Application of a soil development index. *Geoderma* **41** 123–148
[5] Muhs DR 1982 A soil chronosequence on Quaternary marine terraces, San Clemente Island, California. *Geoderma* **28** 257–283.
[6] Aniku JRF, Singer MJ 1990 Pedogenic iron oxide trends in a marine terrace chronosequence. *Soil Sci Soc Am J* **54** 147–152
[7] McFadden LD, Wells SG, Dohrenwend JC 1986 Influences of Quaternary climatic changes on processes of soil development on desert loess deposits of the Cima Volcanic Field, California. *Catena* **13** 361–389
[8] McFadden LD, Wells SG, Brown WJ, Enzel Y 1992 Soil genesis on beach ridges of pluvial Lake Mojave: Implications for Holocene lacustrine and eolian events in the Mojave Desert, southern California. *Catena* **19** 77–79
[9] Reheis MC, Harden JW, McFadden LD, Shroba RR 1989 Development rates of late Quaternary

soils, Silver Lake Playa, California. *Soil Sci Soc Am J* **53** 1127–1140

[10] Harden JW, Taylor EM, Hill C, Mark RK, McFadden LD, Reheis MC, Sowers JM, Wells, SG 1991 Rates of soil development from four soil chronosequences in the southern Great Basin. *Quat Res* **35 (3)** 383–399

[11] Alonso P, Sierra C, Ortega E, Dorronsoro C 1994 Soil development indices of soils developed on fluvial terraces (Penaranda de Bracamonte, Salamanca, Spain). *Catena* **23** 295–308

[12] Dorronsoro C, Alonso P 1994 Chronosequence in Almar River, fluvial terrace soil. *Soil Sci Soc Am J* **58 (3)** 910-925

[13] Scarciglia F, Pulice I, Robustelli G, Vecchio G 2006 Soil chronosequences on Quaternary marine terraces along the northwestern coast of Calabria (Southern Italy). *Quat. Int.* **156/157** 133–155

[14] Singleton GA, Lavkulich LM 1987 A soil chronosequence on beach sands, Vancouver Island, British Columbia. *Can J Soil Sci* **67 (4)** 795–810

[15] Barrett LR, Schaetzl RJ 1992 An examination of podzolization near Lake Michigan using chronofunctions *Can J Soil Sci* **72 (4)** 527–541

[16] Barrett LR 2001 A strand plain soil development sequence in Northern Michigan, USA. *Catena* **44** 163–186

[17] Alexander EB, Burt R 1996 Soil development on moraines of Mendenhall Glacier, southeast Alaska. 1. The moraines and soil morphology. Geoderma 72 1–17

[18] Burt R, Alexander EB 1996 Soil development on moraines of Mendenhall Glacier, southeast Alaska. 2. Chemical transformations and soil micromorphology. Geoderma **72** (1-2) 19–36

[19] Arduino E, Barberis E, Ajmone Marsan F, Zanini E, Franchini M 1986 Iron oxides and clay minerals within profiles as indicators of soil age in northern Italy. *Geoderma* **37** 45–55

[20] Bain DC, Mellor A, Robertson-Rintoul MSE, Buckland ST 1993 Variations in weathering processes and rates with time in a chronosequence of soils from Glen Feshie, Scotland. *Geoderma* **57** 275–293

[21] Mokma DL, Yli-Halla M, Lindqvist K 2004 Podzol formation in sandy soils of Finland. *Geoderma* **120** 259–272

[22] Birkeland PW, Burke RM 1988 Soil catena chronosequences on eastern Sierra Nevada moraines, California, USA. *Arctic and Alpine Research* **20 (4)** 473–484

[23] Egli M, Fitze P, Mirabella A 2001a Weathering and evolution of soils formed on granitic, glacial deposits: results from chronosequences of Swiss alpine environments. *Catena* **45** 19–47

[24] Egli M, Mirabella A, Fitze P 2001b Clay mineral formation in soils of two different chronosequences in the Swiss Alps. *Geoderma* **104 (1-2)** 145-175

[25] Egli M, Mirabella A, Fitze P 2003 Formation rates of smectites derived from two Holocene chronosequences in the Swiss Alps. *Geoderma* **117 (1-2)** 81-98

[26] Pillans B 1997 Soil development at snail's pace: Evidence from a 6 Ma soil chronosequence on basalt in north Queensland, Australia. *Geoderma* **80 (1-2)** 117–128

[27] Muhs DR 2001 Evolution of soils on quaternary reef terraces of Barbados, West Indies. *Quat Res* **56 (1)** 66–78

[28] Bockheim JG 1990 Soil development rates in the Transantarctic Mountains. *Geoderma* **47** (1-2) 59–77

[29] Richter D deB, Yaalon DH 2012 „The changing model of soil" revisited. *Soil Sci Soc Am J* **76** 766–778

[30] Schülli-Maurer I, Sauer D, Stahr K, Sperstad R, Sørensen R 2007 Soil formation in marine sediments of S-Norway: investigation of soil chronosequences in the Oslofjord region. *Revista Mexicana de Ciencias Geologicas* **24** 237–246

[31] Sauer D, Schülli-Maurer I, Sperstad R, Sørensen R, Stahr K 2008 Podzol development with time in sandy beach deposits in southern Norway. *J Plant Nutr Soil Sci* **171** 483–497

[32] Sauer D, Schülli-Maurer I, Sperstad, R, Sørensen, R, Stahr, K, 2009. Albeluvisol development

with time in loamy marine sediments of S-Norway. *Quat Int* **209 (1-2)** 31-43

[33] Sauer D, Finke P, Sørensen R, Sperstad R, Schülli-Maurer I, Høeg H, Stahr K 2012 Testing a soil development model against southern Norway soil chronosequences. *Quat Int* **265** 18-31.

[34] Sauer D, Schülli-Maurer I, Sperstad R, Sørensen R 2013 Micromorphological characteristics reflecting soil-forming processes during Albeluvisol development in S Norway. *Spanish J Soil Sci* **3 (2)** 38-58

[35] Brückner H 1980 Marine Terrassen in Süditalien. Eine quartärmorphologische Studie über das Küstentiefland von Metapont. *Düsseldorfer Geogr Schriften* **14**, 235 pp

[36] Sauer D, Wagner S, Brückner H, Scarciglia F, Mastronuzzi G, Stahr K 2010 Soil development on marine terraces near Metaponto (Gulf of Taranto, southern Italy). *Quat Int* **222**, 48 63

[37] Sauer D, Schellmann G, Stahr K 2007 A soil chronosequence in the semi-arid environment of Patagonia (Argentina). *Catena* **71** 382–393

[38] FAO 2006 *Guidelines for Soil Description.* Fourth edition, FAO, Rome, 97 pp

[39] IUSS Working Group WRB 2014 World Reference Base for Soil Resources. *World Soil Resources Reports* **106**. FAO, Rome, 181 pp

[40] Zander AM, Fülling A, Brückner H, Mastronuzzi G 2006 OSL dating of upper Pleistocene littoral sediments: a contribution to the chronostratigraphy of raised marine terraces bordering the Gulf of Taranto, South Italy. *Geografia Fisica e Dinamica Quaternaria* **29** 33-50

[41] Allen JRM, Brandt U, Brauer A, Hubberten HW, Huntley B, Keller J, Kraml M, Mackensen A, Mingram J, Negendank JFW, Nowaczyk NR, Oberhänsli H, Watts WA, Wulf S, Zolitschka B 1999 Rapid environmental changes in southern Europe during the last glacial period. *Nature* **400 (6746)** 740–743

[42] Brauer A, Allen JRM, Mingram J, Dulski P, Wulf S, Huntley B 2007 Evidence for last interglacial chronology and environmental change from Southern Europe. *PNAS* **104 (2)** 450–455

[43] Hotchkiss S, Vitousek PM, Chadwick, OA, Price J 2000 Climate cycles, geomorphological change, and the interpretation of soil and ecosystem development. *Ecosystems* **3 (6)** 522–533

[44] Johnson DL 1985 Soil thickness processes. In: Jungerius PD (Ed.), *Soils and Geomorphology* **6** Catena Supplement, Braunschweig, 29–40

[45] Johnson DL, Watson-Stegner D 1987 Evolution model of pedogenesis. *Soil Science* **143 (5)** 349-366

Soil in the Anthropocene

Daniel deB Richter[1], Allan R Bacon[2], Zachary Brecheisen[1] and Megan L Mobley[3]
[1] Duke University, Durham, NC USA;
[2] University of Florida, Gainesville, FL USA
[3]University of Wyoming, Laramie, WY USA

E-mail: drichter@duke.edu

Abstract. With scholars deliberating a new name for our geologic epoch, i.e., the Anthropocene, soil scientists whether biologists, chemists, or physicists are documenting significant changes accruing in a majority of Earth's soils. Such global soil changes interact with the atmosphere, biosphere, hydrosphere, and lithosphere (i.e., Earth's Critical Zone), and these developments are significantly impacting the Earth's stratigraphic record as well. In effect, soil scientists study such global soil changes in a science of anthropedology, which leads directly to the need to transform pedostratigraphy into an anthro-pedostratigraphy, a science that explores how global soil change alters Earth's litho-, bio-, and chemostratigraphy. These developments reinforce perspectives that the planet is indeed crossing into the Anthropocene.

1. Introduction

While the sciences of pedology and in fact of all Earth's surface systems have historically focused on *natural* processes, human beings are transforming Earth's surface systems from natural bodies to those that are human-natural bodies. Thus pedology and all Earth system sciences are in the midst of fundamental change.

Human-affected changes in pedology had been written about for decades [1-3], but it was the publication of a small book, *Global Soil Change* [4], that threw open the doors of pedology to human beings as soil-forming agents. No longer could human beings be treated only as agents of soil disturbance. This remarkable book outlined a science of anthropedology as it gave soil science a firm footing as a core environmental science [5].

Today, with over half of humanity living in urban centers, increasingly removed from traditional connections to land and soil, soil and all Earth scientists seem doubly challenged. Not only must the dynamics of global change be scientifically understood and quantified but the increasingly urban public needs continual reminding that the quality of human life and that of the environment is entirely dependent on the Earth's soils and Critical Zone, i.e., the integrated system from the atmosphere and plant boundary layer down through soil into aquifers and the deepest penetration of biogeochemical alteration [6].

This paper' objectives are to consider two interrelated and on-going changes in soil science:

- The transformation of pedology into anthropedology and of soil into a polygenetic and far deeper system than considered in the past. These changes grant to soil science important scientific, intellectual, and societal opportunities.
- The transformation of pedostratigraphy into a science of anthro-pedostratigraphy, which has important implications for the Anthropocene.

2. Pedology Past and Present: Achievements and Developments

The initial scientific conception of soil as a natural body came in the 19th century, that is, soil is a dynamic body that develops structures and processes over time in response to external environmental factors. Three scientists who are given most credit for describing soil's natural forming factors include E.W. Hilgard [7] in *Report on the Geology and Agriculture of the State of Mississippi*, Charles Darwin [8] in *The Formation of Vegetable Mould, through the Action of Worms*, and Vasily V. Dokuchaev [9] in *Russian Chernozem*. Each described soil to be a dynamic system responsive to *natural* environmental forcings. Although Hilgard was employed as a state geologist to survey soils for potential economic development, he carefully addressed his essay entitled, "What is a soil?" by explicitly examining "virgin" rather than human-altered soil. Dokuchaev similarly was employed mapping soils for practical purposes of land taxation, and yet was most influential in his identification of five *natural* soil-forming factors. Darwin was motivated by his own curiosities but after a lifetime of earthworm observation and experimentation, developed an advanced view of the biogenic formation of soil organic matter, i.e., vegetable mould. All three were interested in soil as a natural system; human beings were agents of soil disturbance and were simply not part of pedological studies. Most pedologists have followed the direction set by Hilgard, Darwin, and Dokuchaev, for well over 100 years [5], although three important developments seem important to changes to come.

2.1 Integrating Human Beings into Pedology

Comfortably fitting human beings into the natural body model of soil formation has not been easy [4, 5, 10-14]. Dudal [15] pointedly asked, are we a soil-forming factor short?, out of concern that in a world where over half of Earth's surface was actively managed, a pedogenetic model confined only to natural forming factors risks irrelevance. One of the earliest proposals to more formally integrate human beings into pedology came from Yaalon and Yaron [1]. Amundson and Jenny [10] examined the variety of ways that human beings interacted with soils, in ways that were culturally dependent, potentially self-aware, and modifiable based on human judgment.

One of the main claims of this paper is that the explicit integration of human beings into the natural body model of soil formation, i.e., into anthropedology, provides new opportunities for soil science to explore critical issues of societal relevance. Contemporary pedology can contribute to many environmental programs that continue to grow at many of the world's universities, in governments, and in the private sector. The purview of contemporary pedology includes not only the patterns and processes of natural soil formation but also all the human-forced changes in soils from fertilizers, wastes, biocides, and chemical contaminants; harvests of biomass at various intensities; mixing by cultivation (anthroturbation); alterations from residential, industrial, and transportation uses; biogeochemistry due to drainage or intentional flooding; burning or not burning of organic detritus; alterations due to changing climate; and finally soil sealing and even soil extinction [13, 16]. A remarkable fact is that so much soil science is being practiced today, and not only by soil scientists, but by engineers, environmental chemists, ecologists, biologists, and geoscientists. Although pedologists have unique perspectives and expertise that can frame human-soil interactions within the broader context of soil formation, contemporary pedology risks being incomplete and soon irrelevant if it does not vigorously expand its purview into the non-virgin landscape.

While pedology has always been interdisciplinary, this historic interdisciplinarity has largely been within the natural sciences of geology, biology, chemistry, physics, meteorology, climatology, hydrology, and geochemistry, a list that is impressively long. But as contemporary soils are a human-natural bodies, the list of sciences pertinent to pedology becomes much longer still and far wider ranging. Contemporary pedology involves not only social sciences but also the humanities; the science welcomes insights from economics, sociology, anthropology, and archaeology, and also environmental history, literature, theology, ethics, and even philosophy. Contemporary pedology is a natural and applied science.

2.2 Research Extends the Lower Boundary of Soil

Extending the lower boundary of soil represents a second important change in the traditional model of soil as soil has been greatly expanded in its material volume and process domain. While pedologists in the 19th and 20th centuries increased their depth of interest from O and A into B and C horizons [17, 18], soil studies even today continue to be mainly concentrated in surficial layers (i.e., the solum of the O, A, and B horizons, i.e., "true soil" in the words of Plaster [19]. Even today, many pedologists consider C horizons as "*not* formed by pedogenetic processes."

Here, the practice of how deeply the scientific community samples soil is instructive. Figure 1 illustrates how in over 360 studies of how land-use change alters soil organic carbon, 90% of the 360 studies sampled soil to a maximum of 30-cm depth [20, 21]. While it is often difficult to sample soil, these relatively shallow depths have more to do with how scientists conceive of the soil itself than about sampling difficulty. Despite the fact that plant roots are well known to deeply explore soils to many meters in depth, both in concept and in practice, soil systems are too often treated as if the whole soil profile behaves like the surficial A horizons.

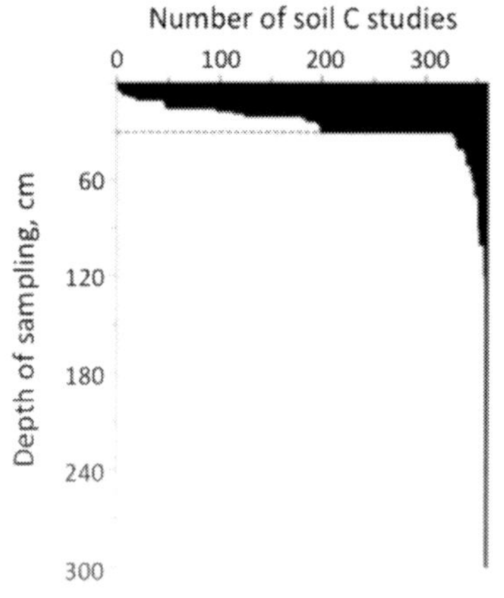

Figure 1. A histogram of the depth of sampling in the 360 studies reviewed by Post and Kwon [20] and West and Post [21] that investigated the effects of land-use change on soil organic carbon. The median depth of sampling was slightly over 15 cm, whereas 90% of the studies sampled soils to 30 cm or less.

In contrast, a concept of the deep soil profile has been advanced by geologists with the so-called weathering profile in which lower layers are described as regolith [22]. While some propose a hybridized pedological-geological profile to avoid conflict in concepts [23], Richter and Yaalon [14] and Richter and Markewitz [18, 24] argue that contemporary understanding of soil requires that C horizons be recognized as fully integral parts of soil profiles. In point of fact, many of processes that form C horizons are fundamentally pedologic. The deep pedologic processes affecting C horizons include soil respiration coupled with carbonic acid weathering, rhizosphere interactions, and pedological alterations of O_2 and redox potential [14, 18, 25-27].

The seasonality of soil respiration and the consequent wave of high CO_2 concentrations result from the annual pulsing of ecosystem metabolism (Figure 2). While most soil CO_2 effluxes back to the aboveground atmosphere, a small but critically important fraction diffuses downward, potentially well below roots and mycorrhizal hyphae, where at least some of it is consumed by weathering [28]. While the CO_2 dynamics of the upper soil layers are well studied [18, 29], the concentration gradient of CO_2 across the lower boundaries of the soil are rarely studied, even though concentration gradients must exist all the way down to the deep CO_2 sink at sites of carbonic acid weathering. In the example of Figure 2, carbonic acid weathering may well be most active at >15 meters in this soil (i.e., weathering profile) in which the full depth of soil plus unconsolidated saprolite (i.e., regolith) is about 30-m over unweathered granite [30].

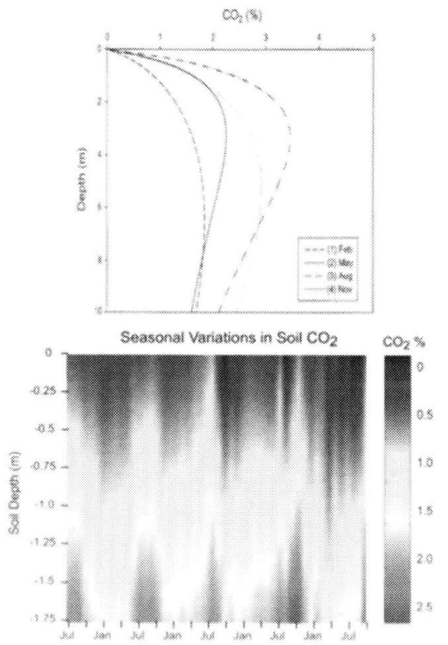

Figure 2. CO_2 at Calhoun Experimental Forest in South Carolina USA. Topmost figure is modeled CO_2 [28] from 0 to 10 m illustrating the pronounced seasonal dynamics of CO_2 concentrations. The figure on the bottom is a Matlab image of CO_2 concentrations measured every two to three weeks over 4.5 years from multiple soil depths.

The lower boundary of soil is not only open, it may be highly heterogenous in chemical and physical structure, diffuse, and temporally variable as well. An organized sequence of depth-dependent biogeochemical fronts of weathering (carbonate and sulfide fronts) was recently described by Brantley et al. [31] in soils and weathering profiles derived from shale. Far more studies of the lower soil boundaries and weathering fronts are needed in a variety of soils and geologic materials. More use of quarries [32] and deep drilling [5] will eventually reveal the variety of biogeochemical and structural patterns in the lower soil boundary (Figure 3).

Finally, the soil's lower boundary conditions are not only under-explored, they are significant for understanding and managing water quality and the sustainability of ecosystems and the environment. With more attention paid to the lower boundary of soil, deep contamination problems can be much better quantified. A notable example is the currently controversial fracking issue and the effects of such mining techniques on the solid, liquid, and gas phases of the full soil system. Considering that the openness of soils and ecosystems is too rarely critically investigated with care and that many lower boundaries extend meters even to many meters, future research of soils' lower boundary conditions and their relations to ecosystem resilience, represents an exciting and significant frontier for pedology and weathering science in general.

2.3 Soil as a Non-Equilibrium, Polygenetic System

A third development in pedology that has moved soil science far from the concepts of even mid-20th century, is a much fuller appreciation for soil-system dynamism, non-equilibrium, and how soils evolve and change over their residence times of 10s to 10^6s of years. A number of scientific disciplines contribute to the understanding that soil-forming factors vary through time, and specifically that soil inputs, removals, transformations, and translocations all vary over the life of most soils (to use the model of Simonson [33]). Such variations can be abrupt as in the periodic phenomena of Butler [34], others more gradual; some changes are externally forced, others derive from internal soil-dependent process. Such system behavior is known as polygenesis, i.e., that the high-order interactions of the factors of soil formation vary significantly over time. How remarkable that within a few decades, the science of pedology has grown from conceiving of soil bodies as largely monogenetic in their formation to understanding that most are polygenetic [35].

Figure 3. Quarries and deep coring are revealing new insights into the soil, saprolite, and the full weathering profile, i.e., into the belowground Critical Zone [32]. Illustrated is a quarry of Carolina slate, a meta-volcantic rock that in the Carolinas yields deep, highly weathered soils and saprolites.

IOP Conf. Series: Earth and Environmental Science **25** (2015) 012010 doi:10.1088/1755-1315/25/1/012010

There are also two ways that soils are polygenetic. One refers to the soil system in place, as profile and soilscape. Most soil profiles have long enough residence times to experience widely variable climates, atmospheric depositions, biotic rooting and organic matter inputs, erosion rates, geomorphic processes, chemical leaching, and human activities. The second meaning of polygenesis refers to the fact that many soil profiles and soilscapes are composed of primary particles that have experienced pedogenesis and weathering when the particles were part of previous soil profiles in the distant past. This latter polygenesis can exhaust weatherable minerals from mineral substrates and contribute much to the understanding of why soils in some regions are largely of advanced weathering stage, i.e., Ultisols and Oxisols.

Accordingly, all soils and their structures and processes are fundamentally diagenetic, formed and changing over time. Pore networks, redoximorphic features, rooting channels, clay mineral suites, nutrient and element distributions, and structural aggregates, all respond to the ebbs and flows of changing environments and biota. Soils are therefore historical bodies, archives of accumulated and inherited features from past conditions of formation -- some features persistent, others disappearing, and others completely erased. Such polygenesis suggests the powerful metaphor of the soil is the palimpsest, a parchment made from animal hide used in the ancient world to record religious communiqués, learned texts, and political decrees, which was written upon, erased when the text was no longer needed, only to be over-written once again. Understanding soils as polygenetic palimpsests requires much more understanding about paleo-climates, biota, geomorphology, changes in geologic substrates, and soil residence times. While we can read the basics of many soils, we are far from understanding soil as an archive, and must realize that at some level the soil may always remain at least a bit shrouded in mystery.

3. Pedology's Future: Anthropocene Soils and a New Science of Anthro-pedostratigraphy

The transformation of Earth's soil from a natural to a human-natural system has many implications. Pedology *must* and will be framed by the human and natural forcings that are hybridizing contemporary soils. In other papers, we argue that human beings' alteration and dependence on soils necessitates long-term soil-ecosystem experiments (LTSEs) that reach their potential if they are passed from one generation of scientists to the next [36, 37, 38]. Here, in addition to the need for LTSEs, we consider a second implication of soil's transition from a natural to human-natural body – that pedostratigraphy be transitioned into a science of anthro-pedostratigraphy. We describe the beginnings of this transition and make the case that anthro-pedostratigraphy will be a useful and special contribution to geologists' deliberations about how the planet has crossed a boundary into the Anthropocene [39, 40].

3.1 Pedostratigraphy

In the past, the science of pedostratigraphy uses paleosols (i.e., technically, soils of the past that are buried, fossilized, and found in the stratigraphic record) to interpret past climates and environments. Geologists have recognized paleosols to be a part of the geologic-climatic record for over a century [41]. In 1961 the American Commission on Stratigraphic Nomenclature instituted formal recognition of paleosols as soil-stratigraphic units. These were not often used but in 1983 the stratigraphic code was revised [42] in part to expand the concept and application of soil-stratigraphic units. Such pedostratigraphic units were named geosols [42, 43] and described to be buried, traceable, three-dimensional bodies of rock that consist of one or more differentiated pedologic horizons with uppermost and lowermost boundaries being the definite physical boundaries of a buried soil profile.

Pedostratigraphy, geosols, and pedo-stratigraphic criteria are summarized well in the latest revision of the North American Stratigraphic Code [44]. Revisions to the code are intended to be slow in coming but they must keep pace with scientific advances. In the 1983 Code, for example, pedostratigraphic strata and geosols were confined to the archaic *solum* concept, i.e., the A, E, and B horizons, excuding the C horizon as "non-soil". In the 2005 revision, however, the C horizon is included in the concept of geosol. According to the 2005 Code, the problem is locating the lower boundary of the C horizon. To the pedologist, however, this is not at all surprising as the lower boundaries of the B and C horizons are frequently the most difficult boundaries to locate in an actively forming soil, much less in a geosol. We hypothesize that ancient, weathered "paleosol C horizons" are common throughout the world but simply have not been widely recognized to date.

While the 2005 Code was modified to expand with the lower boundary of paleosols, understanding and description of new soil orders, notably Andisols, Gelisols, and Oxisols, need consideration in future revisions of the stratigraphic code. Geosols must apparently be overlain by a formally defined lithostratigraphic unit, and thus geosols are rarely if ever found in Holocene materials. Perhaps more clarity is needed on distinguishing pedostratigraphic from other stratigraphic units, and may in this regard be similar to what has been called the archaeosphere [40]. The fundamental core distinction of pedostratigraphic layers is that they are time transgressive, non-synchronous, ie., they only form over time. Paleosols and pedostratigraphic units are often layered in horizons but these cannot be separated from the whole profile, for soil horizons are separated in space but not in time [45]. We suggest that the North American Code and International Stratigraphic Guide [46] be revised on a decadal rather than multidecadal cycle.

3.2 Onward to an anthro-pedostratigraphy

Global soil change indicates how the Earth as a natural planet is transitioning to a human-natural system. A polygenetic wave of human-affected soil change is being recorded in Earth's litho-, bio-, and chemo-stratigraphies and these anthropic signals have yet to be described in the North American Stratigraphic Code and International Stratigraphic Guide. These records form the basis for identifying anthropic strata and boundaries, and they signal how the planet is transitioning from a natural to human-natural system, and crossing into the Anthropocene [39, 40]. Here we sketch some litho-, bio-, and chemostratigraphic signals that record the human alteration of Earth's soil.

3.2.1. Lithostratigraphic signals. Human alteration of soils and paleosols includes soil profiles eroded, mixed with, and buried under new layers of natural and human-made deposits. These "event beds" are the lithostratigraphic signals of the Anthropocene. Most stream and river sediments derive from the world's 1.5×10^9 ha of cultivated soils, although major amounts come from residential construction areas, minelands, roads, kand abandoned, historically cultivated lands. Hooke [47] estimated that humans are Earth's chief geomorphologic agent as gaged by the volume and mass of soil and rock moved by human beings; Wilkinson et al. [48] reinforced this view by estimating that human accelerated rates of sedimentation are an order of magnitude greater than those over geologic time; and Haff [49] expanded on these estimates to include a number of natural and human forced advective and diffusive processes. At the same time, many rivers are dammed and are trapping enormous volumes of eroded soil [50]. Coastlines, always highly dynamic environments, are also actively being modified, supplemented, and restructured in human attempts at stabilization and reclamation [51]. Off-shore, rivers are delivering human-accelerated sediments to the coastal shelf in plumes and large fans of what some might call "subaqueous soil-sediments."

Human beings also create built environments in cities, towns, and villages. The structural materials of such environments are designed to resist degradation, physical and chemical weakening, corrosion, fire, and decomposition. Such structures and their waste products derive from wood, stone, cements, sand, clays, gravels, limestones, mine waste rock, plastics, glass, metals and their alloys. The lower boundary of urban soils are important to consider as the built environment is highly altered belowground. Foundations, pipelines, pilings, and drill boreholes proliferate through the urban subsurface. In the past, with pedology confined as a basic, natural science, the science of urban soils lay practically undeveloped. Today, the pedology of the urban and human-altered environment is a scientific frontier and a burgeoning science [52-54].

3.2.2. Biostratigraphic signals. Because land-use change significantly impacts soil microbes, invertebrates, and vertebrates [55-57], the fact that more than half of Earth's soil are today actively managed means that the Anthropocene is marked by significant impacts on the Earth's biota. While most of the planet's biodiversity resides in the soil, soil's keystone species include engineers such as earthworms, beavers, ants, and mangroves and all are particularly responsive to human activities. Managed and domesticated plants such as corn, rice, and wheat, and perennial crops of trees and vines have root structures and functions that alter soil hydrology, the development of soil horizons, and the physical, biological, and chemical properties of soil. Soil management also alters seed banks, pollen deposits, and the biologically mediated mixing of soil, known as bioturbation.

Human beings have long been responsible for biotic extinctions [58] but also intercontinental mixing of biotic species, which results in some becoming actively invasive. Human-associated extinction of large animals (the mega-fauna) and their replacement by large domesticated cattle, pigs, sheep, and chickens have caused major shifts in the bone records of Pleistocene, Holocene, and Anthropocene soils. Overall, the biostratigraphic signals of human influence are highly significant to the anthro-stratigraphic record.

3.2.3. Chemostratigraphic signals. The biogeochemical cycles of carbon, nitrogen, phosphorus, acidity, metals, radionuclides, and organic compounds have been significantly altered by human beings, and such chemostratigraphic signals in soils and paleosols are global in scale [59, 60]. Nearly all soil profiles and sediments are chemically altered in some way by human imprints from fertilizers, wastes, irrigation solutes, and various airborne pollutants and radionuclides.

Chemical markers define many geologic boundaries, one of the most famous being the iridium anomaly of the Cretaceous-Paleogene boundary when the extraterrestrial bolide collided with the Earth approximately 65 million years ago. Isotope excursions of several elements are also used to mark geological boundaries and may be particularly useful to signaling the Anthropocene given the human emissions to the atmosphere of a family of long-lived isotopes. Lead isotopes demonstrate pollution histories over several millennia of mining and smelting [61], and sulfur isotopes trace sources of acid rain from the industrial age. Stable isotopes of carbon and nitrogen track effects of fossil fuel combustion and the rapid rise of nitrogen fertilization in the late 20th century. Radioisotopes from aboveground nuclear bomb testing, including ^{99}Tc, ^{129}I, and ^{239}Pu, are detectable across the planet due to their liberation and long-distance transport from the 1950s through the mid-1960s [62].

4. Conclusions: An Anthro-pedostratigraphy for the Anthropocene

Over much of the Holocene, human beings have fundamentally transformed soil, biologically, chemically, and physically. Paleosol formation has been greatly accelerated as both fluvial and aeolian erosion and deposition have been greatly increased relative to long-term natural rates. As soils are transformed into human-natural systems, soil fluxes of energy, gases, water, solutes, and solids and energy with the atmosphere, hydrosphere, lithosphere, and biosphere are altered in ways difficult to predict. The human imprint on Earth's soil has created the need for a new science, that of anthro-pedostratigraphy, to track the effects of human altered soils and paleosols on litho-, chemo-, and bio-stratigraphies. The transformations of both pedology to anthropedology and pedostratigraphy into anthro-pedostratigraphy signal clearly that a boundary has been crossed: the Earth itself is transitioning to a human-natural system, i.e., in the Anthropocene.

Acknowledgement
This paper was written based on the first author's Russell Memorial Lecture given to the British Society for Soil Science in Lancaster University, September 2013, on a Keynote Lecture at the Soil Change Matters Workshop at The Capital in Bendigo, Australia, March 2014, and on a seminar given to Stanford University's Environmental Earth System Science Department in May 2014. Several research grant and fellowship programs provided funding for this research: National Science Foundation Doctoral Dissertation Improvement Grants DEB-1011186 to DD Richter and ML Mobley and DEB-131614 to DD Richter and AR Bacon, National Science Foundation Ecosystems Grant DEB-0717368 to DD Richter, National Science Foundation CZO Grant EAR-1331846 to DD Richter.

References

[1] Yaalon, DH and B Yaron 1966 Framework for man-made soil changes – an outline of metapedogenesis *Soil Sci* **102** 272-277
[2] Bidwell, OW, and FD Hole Man as a factor of soil formation *Soil Sci* **99** 65-72
[3] Bouma, J and FD Hole 1971 Soil structure and hydraulic conductivity of adjacent virgin and cultivated pedons at two sites A typic argiudoll (silt loam) and a typic eutrochrept (clay) ***Soil Sci Soc Am Pro*** **35** 316-19
[4] Arnold, RW, J Szabolcs, and VO Targulian 1990 *Global Soil Change* International Institute for Applied Systems Analysis, Laxenburg, Austria
[5] Richter, D deB 2007 Humanity's transformation of Earth's soil. Pedology's new frontier Soil *Sci* **172** 957-67
[6] Richter DdeB and SA Billings 2015 "One physical system": Tansley's ecosystem as Earth's critical zone *New Phytol* (in press)
[7] Hilgard, EW 1860 *Report of the Geology and Agriculture of the State of Mississippi* E Barksdale, State Printer, Jackson, MS, USA
[8] Darwin, C R 1881 *The Formation of Vegetable Mould* John Murray, London
[9] Dokuchaev, VV 1883 *Russian Chernozem* In Selected Works of VV Dokuchaev, Vol 1, p 14–419 Moscow, 1948 Israel Program for Scientific Translations Ltd (for USDA-NSF), S Monson, Jerusalem, 1967 (Translated from Russian into English by N Kaner)
[10] Amundson, R and H Jenny 1991 The place of humans in the state factor theory of ecosystems and their soils *Soil Sci* **151** 99-109
[11] Paton, TR, GS Humphreys, and PB Mitchell 1996 *Soils – A New Global View* Yale University Press, New Haven, CT, USA
[12] Schaetzl RJ and ML Thompson 2015 *Soils: Genesis and Geomorphology* Cambridge University Press, UK
[13] Richter, DdeB et al 2011 Human–soil relations are changing rapidly Proposals from SSSA's cross-divisional soil change working group ***Soil Sci*** *Soc Am J* **75** 2079-84
[14] Richter, DdeB and DH Yaalon 2012 "The changing model of soil" revisited *Soil Sci Soc Am J* **76** 766-78
[15] Dudal, R 2005 The sixth factor of soil formation *Eurasian Soil Sci+* **38** S60
[16] Amundson, R, Y Guo, and P Gong 2003 Soil diversity and land use in the United States *Ecosystems* **6** 470-482
[17] Nikiforoff, CC 1959 Reappraisal of the soil *Science* **129** 186-196

[18] Richter, DD and D Markewitz 1995 How deep is soil? *BioScience* **45** 600-09
[19] Plaster, EJ *Soil Sci Manage* Delmar Publishers, Albany, NY, USA
[20] Post, WM, and KC Kwon 2000 Soil carbon sequestration and land-use change processes and potential *Global Change Biol* **6** 317-27
[21] West, TO, and WM Post 2002 Soil organic carbon sequestration rates by tillage and crop rotation *Soil Sci Soc Am J* **66** 1930-46
[22] Pavich, M J, L Brown, JN Valette-Silver, J Klein, and R Middleton 1985 ^{10}Be analysis of a Quaternary weathering profile in the Virginia Piedmont *Geology* **13** 39-41
[23] Tandarich, JP, RG Darmody, LR Follmer, DL Johnson 2002 Historical development of soil and weathering profile concepts from Europe to the United States of America *Soil Sc Soc Am J* **66** 335-46
[24] Richter, DD and D Markewitz 2001 *Understanding Soil Change* Cambridge, UK
[25] Fimmen, RL, D deB Richter, D Vasudevan, MA Williams and LT West 2008 Rhizogenic Fe–C redox cycling a hypothetical biogeochemical mechanism that drives crustal weathering in upland soils *Biogeochemistry* **87** 127-41
[26] Richter, DdeB, NH Oh, RL Fimmen, and JA Jackson, J 2007b The rhizosphere and soil formation pp 179-200 In *The Rhizosphere An Ecological Perspective* ZG Cardon and JL Whitbeck (Ed) Academic Press, Amsterdam
[27] Bacon, AR 2014 *Pedogenesis and Anthropedogenesis on the Southern Piedmont* Dissertation Duke University, Durham, NC, USA
[28] Oh NH, Kim HS, Richter DD. 2005. What regulates soil CO_2 concentrations? A modeling approach to CO_2 diffusion in deep soil profiles *Environ Eng Sci* **22** 38-45
[29] Bernhardt, ES, JJ Barber, JS Pippen, L Taneva, L, JA Andrews, and WH Schlesinger 2006 Long-term effects of free air CO2 enrichment (FACE) on soil respiration *Biogeochemistry* **77** 91-116
[30] Bacon, AR, DeB Richter, PR Bierman, and DH Rood 2012 Coupling meteoric ^{10}Be with pedogenic losses of ^{9}Be to improve soil residence time estimates on an ancient North American interfluve *Geology* **40** 847-50
[31] Brantley, SL, ME Holleran, L Jin, and E Bazilevskaya 2013 Probing deep weathering in the Shale Hills Critical Zone Observatory, Pennsylvania (USA) the hypothesis of nested chemical reaction fronts in the subsurface *Earth Surf Processes* **38** 1280-98
[32] Oh, N-H and DD Richter 2005 Elemental translocation and loss from three highly weathered soil–bedrock profiles in the southeastern United States *Geoderma* **126** 5-25
[33] Simonson, RW 1959 Outline of a generalized theory of soil genesis *Soil Sci Soc Am J* **23** 152-56
[34] Butler, BE 1959 *Periodic Phenomena in Landscapes as a Basis for Soil Studies* CSIRO, Melbourne, Australia
[35] Buol, SA, RJ Southard, RC Graham, and PA McDaniel 2003 Soil genesis and classification Blackwell, Oxford, UK
[36] Ladha, JK, D Dawe, H Pathak, AT Padre, RL Yadav, B Singh, and PR Hobbs 2003 How extensive are yield declines in long-term rice–wheat experiments in Asia? *Field Crop Res* **81** 159-80
[37] Richter, DdeB, M Hofmockel, MA Callaham, DS Powlson, and P Smith 2007a Long-term soil experiments Keys to managing Earth's rapidly changing ecosystems *Soil Sci Soc Am J* **71** 266-79
[38] Jenkinson DS, PR Poulton, C Bryant C 2008 The turnover of organic carbon in subsoils Part 1 Natural and bomb radiocarbon in soil profiles from the Rothamsted long-term field experiments *Eu J Soil Sci* **59** 391-99
[39] Zalasiewicz, J, et al 2015 When did the Anthropocene begin? A mid-twentieth century boundary level is stratigraphically optimal *Quaternary International* DOI: 10.1016/j.quaint.2014.11.045

[40] Edgeworth, M, D deB Richter, CN Waters, P Haff, C Neal, SJ Price Diachronous beginnings of the Anthropocene the stratigraphic bounding surface between anthropogenic and non-anthropogenic deposits *Anthropocene Review* DOI: 10.1177/2053019614565394
[41] Leverett, F 1898 The weathered zone (Yarmouth) between the Illinoian and Kansas till sheets *J Geol* 6 238-43
[42] North American Commission on Stratigraphic Nomenclature (NACSN) 1983 North American Stratigraphic Code *Am Assoc Petr Geol B* **67** 841-75
[43] Morrison, RB 1967 Principles of Quaternary soil stratigraphy *In* RB Morrison and HE Wright, Jr (eds) *Quaternary Soils* Center for Water Resources Research, Desert Research Institute, University of Nevada, Reno, USA
[44] North American Commission on Stratigraphic Nomenclature (NACSN) 2005 North American Stratigraphic Code *Am Assoc Petr Geol B* **89** 1547-91
[45] Cremeens, DL and JP Hart 1995 On chronostratigraphy, pedostratigraphy, and archaeological context In *Pedological Perspectives in Archaeological Research* Soil Science Society of American Special Publications 44 Madison, Wisconsin, USA
[46] Salvador, A (ed) 1994 *International Stratigraphic Guide* International Union of Geological Sciences, London
[47] Hooke, R LeB 2000 On the history of humans as geomorphic agents *Geology* **28** 843-46
[48] Wilkinson, BH and BJ McElroy 2007 The impact of humans on continental erosion and sedimentation *Geol Soc Am Bull* **119** 140-56
[49] Haff, PK 2010 Hillslopes, rivers, plows, and trucks: mass transport on Earth's surface by natural and technological processes *Earth Surf Processes* **35** 1157-66
[50] Walling, DE 2009 *The impact of global change on erosion and sediment transport by rivers current progress and future challenges* UNESCO, Paris
[51] Burke, L, Y Kura, K Kassem, C Revenga, M Spalding, and D McAllister 2001 Pilot Analysis of Global Ecosystems Coastal Ecosystems World Resources Institute, Washington DC
[52] Pulleman, MM, J Bouma, EA van Essen and EW Meijles 2000 Soil organic matter content as a function of different land use history *Soil Sci Soc Amer J* **64** 689-94
[53] Sonneveld, MPW, JBouma and AVeldkamp 2002 Refining soil survey information for a Dutch soil series using land use history *Soil Use Manage* **18** 157-63
[54] Lorenz, K and R Lal 2009 Biogeochemical C and N cycles in urban soils *Environ Int* **35** 1-8
[55] Callaham, MA, DD Richter, DC Coleman, and M Hofmockel 2008 Long-term land-use effects on soil invertebrate communities in Southern Piedmont soils, USA *Eu J Soil Biol* **42** S150-56
[56] Coleman, DC, and DA Crossley 1996 *Fundamentals of Soil Ecology* Academic Press, New York
[57] Wardle, DA 2002 *Communities and Ecosystems* Princeton University Press, NJ, USA
[58] Steadman, DW, PS Martin, DE Ross, AJT MacPhee, HG Jull, McDonald, CA Woods, M Iturralde-Vinent, and GWL Hodgins 2005 Asynchronous extinction of late Quaternary sloths on continents and islands *P Natl Acad Sci USA* **102** 11763-68
[59] Schlesinger, WH and ES Bernhardt 2013 *Biogeochemistry* Academic Press, New York
[60] Vitousek PM, JD Aber, RW Howarth, GE Likens, PA Matson, DW Schindler, WH Schlesinger, DG Tilman. 1997. Human alteration of the global nitrogen cycle: sources and consequences. *Ecol Applic* **7** 737-50
[61] Renberg, I, R Bindler, and M-L Brännvall 2001 Using the historical atmospheric lead-deposition record as a chronological marker in sediment deposits in Europe *The Holocene* **11** 511-16
[62] Waters, CN, J Zalasiewicz, M Williams, MA Ellis, A Snelling 2014 *A Stratigraphical Basis for the Anthropocene* London Geological Society, Special Publications 395

IOP Conf. Series: Earth and Environmental Science **25** (2015) 012011 doi:10.1088/1755-1315/25/1/012011

Permanent soil monitoring system as a basic tool for protection of soils and sustainable land use in Slovakia

J Kobza
National Agricultural and Food Centre - Soil Science and Conservation Research Institute Bratislava - Regional working place Banska Bystrica and Faculty of Natural Sciences, Matej Bel University, Banska Bystrica, Slovakia, Europe

E-mail: j.kobza@vupop.sk

Abstract. The purpose of soil monitoring system in Slovakia is to better protect the soils with regard to sustainable land use. The main object is the observation of soil parameters indicative of change to the equilibrium of soil system as far as to the irreversible change with possible development of degradation processes in soil. The soil monitoring system in Slovakia has been running since 1993. Its importance consists of providing the information on changing spatial and temporal variations of soil parameters as well as the evolution of soil quality in topsoil and subsoil. The soil monitoring network in Slovakia is constructed using ecological principles, taking into account all main soil types and subtypes, soil organic matter, climatic regions, emission regions, polluted and non-polluted regions as well as various other land uses. The results of soil monitoring of 318 sites on agricultural land in Slovakia have been presented. Soil properties are evaluated according to the main threats to soil relating to European Commission recommendation for European soil monitoring performance as follows: soil erosion, soil compaction, decline in soil organic matter, soil salinization and sodification and soil contamination. The most significant change has been determined in physical properties of soils. The physical degradation was especially manifested in compacted and the eroded soils. On the basis of our results about 40 % of agricultural land is potentially affected by soil erosion in Slovakia. In addition, decline in soil organic matter and available nutrients indicate seriousness of soil degradation processes observed during the last monitoring period in Slovakia. Measured data and required outputs are reported to Joint Research Centre (JRC) in Ispra (Italy) and European Environmental Agency (EEA) in Copenhagen (Denmark). Finally, the soil monitoring system thus becomes a basic tool for protection of soils and sustainable land use as well as for the creation of legislation not only in Slovakia, but in EU, too.

1. Introduction

Conception of the European soil policy and soil protection strategy as well as its sustainable use of land was established in the 6th Environmental Action Programme, which was accepted by European Council as well as European Parliament on the 22nd of July, 2002. The basic strategy involved developing a system to monitor soil and changes in soil over time. The main aim of the soil monitoring system is to obtain the data of the most current soil condition and to monitor changes in soil properties over time according to concrete threats to soil (soil contamination, salinization and sodification of soils, decline in soil organic matter, soil compaction and erosion) in Slovakia, what is in harmony with the European strategy of soil monitoring there are permanently monitored important parameters in connection to recommendations of EC for evaluation of actual state and development of soils [1].

Government soil policy in Slovakia declares that the soil is and will be the basic building block of environmental, ecological, economic and social potential of Slovakia and therefore it must be carefully protected against damage. The new regulation concerning agricultural soils are Act n. 220/2004 Z.z. on protection of agricultural soils and agricultural land use [2] and Act n. 59 of Ministry of Agriculture and Rural Development of Slovakia from the 11.-th of March, 2013 [3]. The aim of the Acts is to provide protection of agricultural soils against degradation. The soil monitoring system developed in Slovakia is therefore an essential tool to identify the current state of soils and monitor future changes to assist with the effective implementation of these regulations.

The soil monitoring system in Slovakia has been running consistently since 1993. Its importance consists of providing actual and objective information on temporal trends in important soil properties on agricultural soils of Slovakia. Data measured and collected are reported to JRC (Joint Research Centre) in Ispra (Italy) and to EEA (European Environmental Agency) in Copenhagen (Denmark).

2. Material and Methods

The soil monitoring network in Slovakia is constructed on ecological principles and includes research data of all main soil types and subtypes, soil substrates, climatic regions, emission regions, polluted and non-polluted regions as well as various land use (Figure 1). There are 318 monitoring sites on agricultural and alpine land in Slovakia. All soil monitoring sites are located in WGS 84 coordinates. The monitoring site is represented by a circular shape, of radius 10 m and an area of 314 m^2. The standard depths of 0–0.10 m, 0.20–0.30 m and 0.35–0.45 m on soils under grassland and 0–0.10 m and 0.35–0.45 m on arable land are sampled, but these depths are adjusted to correlate with the main soil horizons. The soil monitoring in Slovakia is carried out on 5 year cycles. The most important soil indicators concerning threats to soil are included in the soil monitoring system in Slovakia according to the recommendation of the European Commission (EC) for united soil monitoring system in Europe [1].

Methodical and analytical procedures have been realized according to work [4].

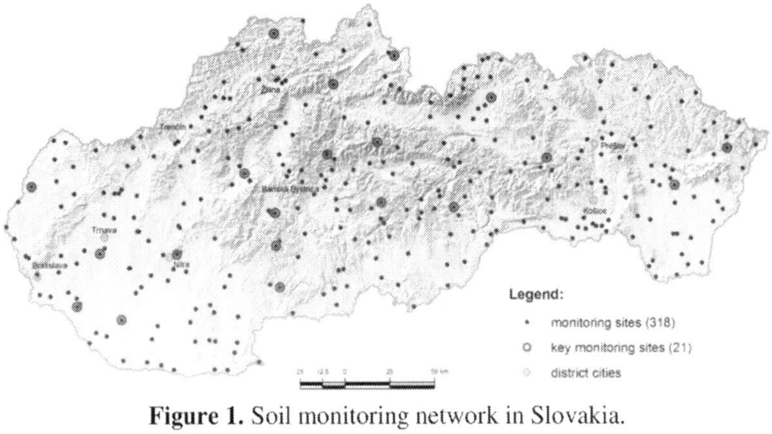

Figure 1. Soil monitoring network in Slovakia.

2.1. Monitored indicators of soil monitoring system according to threats to soil in Slovakia

The most important soil indicators concerning threats to soil, according to Recommendation of European Commission (EC) for unified soil monitoring system in Europe [1], are included in soil monitoring system for Slovakia. These threats and their associated indicators include:

Soil contamination:
- Cd, Cr, Pb, Ni, Zn, Cu, Se, Co (extracted with *aqua regia*), Hg (total content);
- As, Cu, Ni, Zn, Cd, Pb (extracted with 1 mol/l NH_4NO_3) – bioavailable forms;
- Fwater. (watersoluble fluorine measured by ionselective electrode);
- Polycyclic aromatic hydrocarbons (PAH_s);
- Polychlorinated biphenyls (PCB);
- Non-polar extractive substances

Soil acidification:
- pH/H_2O;
- pH/KCl;
- $pH/CaCl_2$;
- cation exchange capacity (CEC);
- exchangeable cations (Ca^{2+}, Mg^{2+}, K^+, Na^+);
- active Al (if pH/KCl<6.0).

Salinization and sodification:
- electrical conductivity (ECe);
- exchangeable sodium percentage (ESP);
- sodium adsorption ratio (SAR);
- pH/H_2O;
- exchangeable cations and anions (Ca^{2+}, Mg^{2+}, K^+, Na^+, Cl^-, SO_4^{2-}, CO_3^{2-}, HCO_3^-).

Quantitative and qualitative composition of soil organic matter:
- organic carbon (Cox);
- total nitrogen (Nt);
- humic acids (HA) and fulvo acids (FA);
- colour quotient (Q^4_6);
- fractional composition of humic acids (elementary analysis of C, H, N, O).

Macro- and micronutrients:
- available macronutrients P, K, Mg (Mehlich III);
- micronutrients Cu, Zn, Mn (DTPA).

Soil compaction:
- bulk density (ςd);
- porosity (P);
- maximum capillary water capacity (wKMK);
- texture (according to FAO).

Soil erosion (on selected soil transects):
- ^{137}Cs;
- pH/KCl;
- Cox;
- P, K;
- texture (according to FAO).

3. Results and discussion

3.1. Soil erosion

Soil erosion is a significant environmental problem and the most widespread degradation process in Slovakia. Erosion is measured on soil transects using ^{137}Cs profile distribution. In addition, the area of soil erosion distribution is determined by using the erosion prediction model that incorporates the USLE [5]. Results showed that there are actually about 40 % of agricultural soils affected by soil erosion in Slovakia (Table 1).

Table 1. Soil erosion distribution in agricultural land of Slovakia.

Degree of erosion **(soil loss)**	**Area, ha**	**% of farmland**
Normal profile and slightly eroded (0–4 t ha^{-1} year^{-1})	1, 454. 925	60.35
Moderately eroded (4–10 t ha^{-1} year^{-1})	245. 421	10.18
Severely eroded (10–30 t ha^{-1} year^{-1})	356.318	14.78
Extremely severely eroded (>30 t ha^{-1} year^{-1})	354. 148	14.69
Total	2, 410. 812	100.00

In addition, an interactive and predictive soil erosion model was created for use by the farmers. The model can be found at http:/www.podnemapy.sk [6]. This application can be useful for providing information concerning soil - erosion intensity on agricultural field using the following procedure: Users of agricultural fields (farmers) move step by step through the model from region level through district, cadaster as far as parcel (field) level where the important data on soil erosion are required (Figures 2a, b). The farmers can obtain the available data on following factors of soil erosion (R – rain factor, K – factor of erodibility, LS – factor of relief influence, C – factor of vegetation cover, P-factor of measures against erosion) and the amount of potential and actual soil loss in interaction with agricultural crop on agricultural field can be calculated.

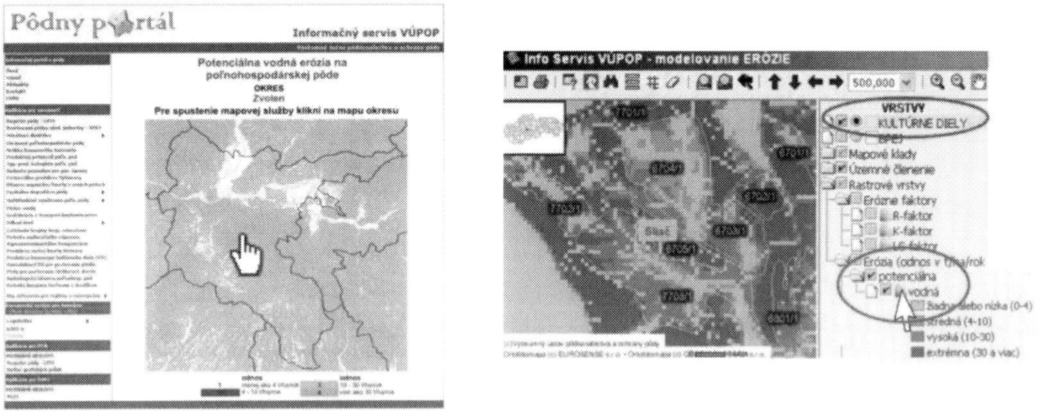

Figures 2 a, b. Soil portal with the image of potential water erosion.

3.2. Soil compaction

Soil compaction is monitored in the soil monitoring network only on arable land (topsoil and subsoil are included) on the basis of measured physical indicators such as bulk density, porosity and texture.

1. Primary compaction – controlled by natural characteristics of the soil (soil texture, soil organic matter content)
2. Secondary compaction - influenced by human activity (kind of machinery used, compliance with optimal soil conditions for machinery use, mainly soil moisture at the time of the agronomical operations).

The main indicator for soil compaction is bulk density. Its values in Slovakian soils are running between 1.0 - 1.8 kg m^{-3}. According to measured bulk density values (Figure 3) the soils sensitive to soil compaction occur mostly in south-western and south-eastern parts of Slovakia (these are the lowlands used mostly for agricultural – arable soils, loamy to clayey textures with usage of heavy machines).

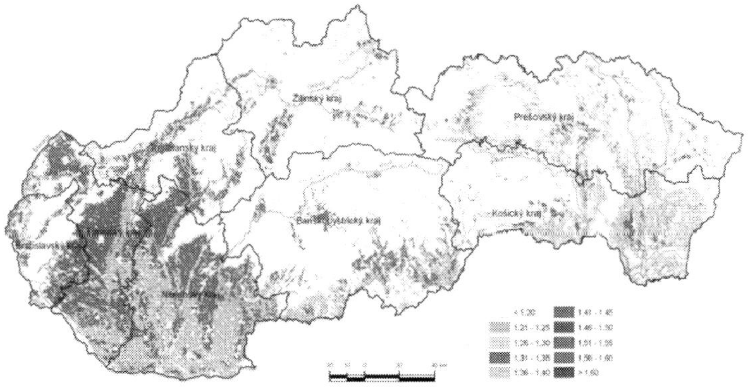

Figure 3. Map of topsoil bulk density in Slovakia (kg .m^{-3}).

Table 2. Percentage of monitoring sites showing compaction according to soil types (average values from 4 previous monitoring cycles).

Soil profile (depth)	**Soil texture**	**ČA**	**ČM**	**FM**	**HM**	**KM**	**PG**
Topsoil (0-0.10 m)	sandy	17	-	-	-	-	-
	loamy	25	35	27	27	28	37
	clayey	45	63	38	31	36	35
Subsoil (0.30-0.35 m)	sandy	67	-	-	-	-	-
	loamy	56	46	59	60	72	82
	clayey	72	67	70	90	81	87

Explanations: ČA – Mollic Fluvisol, ČM – Chernozem, FM – Fluvisol, HM – Luvisol, KM – Cambisol, PG – Planosol

Percentage of monitoring sites showing compaction according to soil types and soil texture are given in table 2. The soil monitoring network in Slovakia consists of 318 monitoring sites (100 %). On the basis of obtained results it may be said that the most severe compaction occur in subsoil especially on clayey soils. Differences in soil compaction between soil types are not very significant. Finally, it may be said that the physical conditions of observed soils are deteriorating in the direction from texturally loamy to clayey soils which are the most sensitive to soil compaction. There are about 200 000 hectares of compacted agricultural soils and about 500 000 hectares potentially compacted agricultural soils in Slovakia.

3.3. Decline in soil organic matter

Soil organic matter represents more than 95 % of total carbon contained in soils under grassland and near 100 % of total carbon accumulated in soils under arable land [7]. Quantitative and qualitative indicators of soil organic matter (SOM) are permanently monitored in soil monitoring network in

Slovakia. After slight decline in soil organic carbon (SOC) at the beginning of soil monitoring system in Slovakia (1993 year), all arable soils showed an increase in SOC its increase has been indicated on all arable soils during the last monitoring period (Figure 4). The increase could be caused by subsidies from the Slovak Government for increasing of soil organic matter in soils.

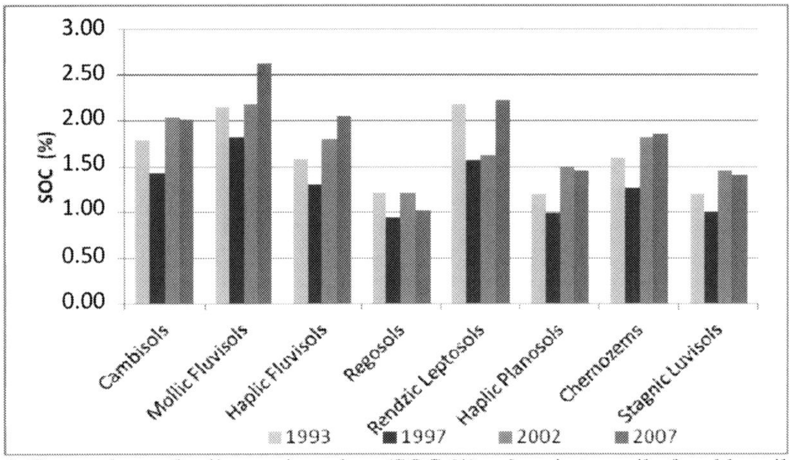

Figure 4. Comparison of soil organic carbon (SOC %) values in topsoil of arable soils of Slovakia.

Qualitative indicators of soil humus (HA/FA ratio, colour quotient Q^4_6, fractional composition of humic acids - HA) are without significant trends. However, the measured values are characteristic for the concrete soil type as well as for the chemical structure of humic acids (HA). These indicators seem to be a result of soil genesis.

The content of organic carbon (SOC) in soils is largely conditional upon the processes of soil formation. The content of SOC in cultivated soils, mainly in arable soils, is limited by intensity and depth of cultivation that affect degree of mineralization. For this reason the average values of SOC in arable Slovak soils ranged from 1 to 2.5 %. This corresponds to the moderately and good content of soil humus when the values of SOC is converted into humus (conversion coefficient is 1.724). The lowest values of soil organic carbon are typical for Regosols and the highest values of SOC – for Mollic Fluvisols.

Recent data indicate that changes in SOC levels in arable soils differ depending on soil type. The gradual increase of SOC concentration is clear mainly in Mollic Fluvisols and Haplic Fluvisols. The average value of SOC in Rendzic Leptosols after an early decline then displayed a substantial back to the original levels SOC values in Chernozems and Cambisols have stabilised at the same level as in the previous cycle. Slight reduction of SOC content in Haplic Planosols and Stagnic Luvisols were observed, but they were not statistically significant. Proportionately, the greatest reduction of SOC level was observed in Regosols.

In summary, after a slight decline in soil humus, especially in arable land, since the beginning of soil monitoring the tendency for stabilization of soil humus content and in some cases an increase in soil humus has been observed during more recent monitoring periods in Slovakia. Changes in total nitrogen are not significant except in some soils such as Cambisols, Chernozems, Mollic and Haplic Fluvisols and Luvisols, where only slight increase has been registered. In qualitative parameters of humus (HA/FA, Q^4_6) no significant changes were indicated during last 20 years (period of soil monitoring in Slovakia).

3.4. Soil salinization and sodification

On the basis of obtained results it was indicated that *salinization and sodification* processes are running more or less in parallel, but the sodification process seems to be dominant under the soil - and climate conditions of Slovakia, and are to be found mostly in south-western and south-eastern parts of

Slovakia [8]. There estimated to be three thousand hectares of saline soils in Slovakia that represents 0.1 % of agricultural land.

Salinization is a process of the accumulation of neutral sodium salts in soil, primarily NaCl and Na_2SO_4 is the indicators of salinization process are the total content of water-soluble salts in soil and electrical conductivity of saturated soil extract (ECe).

Sodification is a process of the fixation of exchangeable sodium in sorption complex of soil. This process is influenced by the presence of alkaline salts in soil, primarily Na_2CO_3, $NaHCO_3$ and Na_2SiO_3. The content of exchangeable sodium in soil sorption complex (ESP) and pH value are indicators of the sodification process.

These processes are running more or less simultaneously, but the sodification process seems to be dominant under conditions of Slovakia.

In principle, it is primary (Figure 5) or secondary, anthropogenic, processes (e.g. strongly alkaline waste deposits from aluminium production - Figure 6).

Figure 5. Example of the primary saline soil located under semiarid conditions.

Figure 6. Example of the secondary saline and polluted soil influenced by strongly alkaline waste deposits.

3.5. Soil contamination

The significant change in concentration of inorganic (Cd, Pb, Cu, Zn, Cr, Ni, Co, Se, As, Hg and F) and organic (PAHs and PCB) contaminants was not observed during the of 20 years monitored period. It appears that the soils which were contaminated at the beginning of soil monitoring, are still contaminated at present. There are about 20 thousand hectares of contamined soils in Slovakia. This constitutes 1 % of the total area of soils in Slovakia, but the predominant part of them is contaminated by geogenic influence.

The various impacts influencing soil contamination can be described as follows:

- human influence (industry, agriculture, municipal waste materials, etc.);
- influence of geochemical anomalies;
- combination of the above.

The human influence on contamination of soils in Slovakia (former Czechoslovakia) was most significant after the Second World War and especially during the industrial period in the second half of the 20-th century. Later, after 1990, soil pollution was slight, without significant changes. It means that soils, which were polluted in the past have remained practically unchanged and continue to be in an unhealthy condition.

The observation of geochemical anomalies occurs particularly in volcanic and crystalline rocks, found mainly in mountainous regions and more rarely within agricultural land. The most extensive areas of geochemical anomalies occur in the Štiavnické vrchy (mountains), Low Tatras and Slovenské Rudohorie (mountains). These regions are characterized by high and very high concentrations of risk trace elements, especially heavy metals (Cd, Pb, Cu, Zn, As and Hg) in all soil profiles.

Finally, the highest concentrations of risk trace elements have been registered in soils in the areas with a combination of anthropogenic and geogenic contamination, in Central Spiš region – Krompachy and Rudňany industrial-metallurgical zone, where the high concentration of risk trace elements, primarily Cd, Pb, Cu, Zn, As and Hg were determined.

The type of soil contamination determines the type of remediation with subsequent land use and soil protection. The basic criterion is to limit transport of contaminants' into the food chain and crops.

It is possible to use several remediation techniques regarding to pollutants and soils exposed to anthropogenic pollution [9, 10].

3.6. Soil monitoring database of Slovakia and data reporting

The database of the soil monitoring system allows to creation and maintenance of data for each of the monitoring sites of agricultural land as well as the preparation of data for further processing through specialized programs (statistical programs, spreadsheets and other databases). Position information provides a link to the geographical information systems, and thus opens the possibilities for the further spatial analysis and representation of results.

User interface database also allows insertion and viewing of data, their selection and printing tables of data stored on the soil profiles.

Structural harmonization of the database, rebuilding the original structure of the geodatabase and the transition to an ORACLE platform are progressing simultaneously. In addition, a new network interface database system also has been built to facilitate new data capture. The main functional services for publishing the spatial information of soil monitoring are on-line with OGC - WMS & WFS specifications to define the standard reporting data structures.

In addition, the database is stored to the regular monitoring data which are archived from a network of monitoring sites where the soil samples have been sampled and analysed in monitoring period. Metadata about performance of monitoring period are attached in the database – together with updated data, descriptions of soil monitoring site characteristics and soil classification.

Figure 7. Integration of soil monitoring database with other soil databases in Slovakia.

The integrated soil database in Slovakia enables value adding of information from the soil monitoring system to be achieved. In figure 7 the application of various digital layers (land use database, ortophotomaps of landscape, LUCAS and soil monitoring database) can better explain actual soil changes caused by different and changed land uses around known monitoring site (400044 in red above).

Ten partial monitoring systems of environment in Slovakia are described in Figure 8. The data summaries are regularly reported to the European Environmental Agency in Copenhagen (Denmark) and Joint Research Centre in Ispra (Italy).

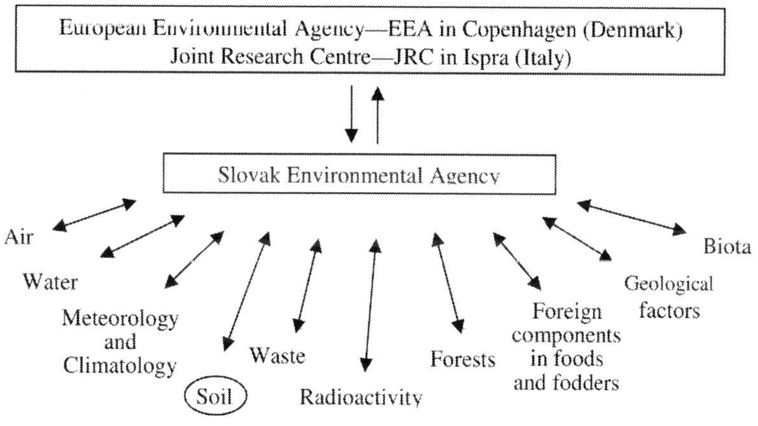

Partial monitoring systems of environment in Slovakia

Figure 8. Implementation of soil monitoring system with monitoring of environment in Slovakia and EU.

4. Conclusions

The soil monitoring system is the source of the comparable and objective data on the current state and evolution of soils in Slovakia. The obtained results are useful in decision-making (e.g. for soil protection) and in various branches of national economy as well as in research institutes and universities with environmental focus. In addition, reporting of stored data to EEA in Copenhagen (Denmark) and to JRC in Ispra (Italy) contributes to important outputs concerning the actual state and evolution of soil cover in EU.

The most significant change has been recorded in physical properties of soils. The physical degradation was especially manifested in compacted and the eroded soils. About 40 % of the agricultural land is potentially affected by soil erosion in Slovakia. In addition, the decline in soil organic matter and available nutrients (mostly P and K) are evidence of continued existing soil degradation processes observed during the last monitoring period in Slovakia.

Finally, the soil monitoring system has become a basic tool for protection of soils and sustainable land use as well as for creation of legislation not only in Slovakia, but in EU, also. The role of soil monitoring system in the protection of soil and land use could be identified as follows:

- identification of risk areas and their assessment;
- prognosis of the subsequent evolution of soil parameters according to threats to soil;
- a basic tool for protection of soils and sustainable land use;
- creation of ecological legislation;
- evaluation of soils and environment (including recommendations and measures).

Acknowledgements
We would like kindly to thank Slovak Ministry of Agriculture and Rural Development for the financial support of this project.

References

[1] Van–Camp L, Bujarrabal B, Gentile A-R, Jones R J A, Montanarella L, C. Olazabal C, and Selvaradjou S-K 2004 *Reports of the Technical Working Groups Established under the Thematic Strategy for Soil Protection*, EUR 21319 EN/5 (Luxembourg: Office for Official Publications of the European Communities) p. 872

[2] Ministry for Land Management 2004 *Act No. 220/2004 on protection and agricultural land use. Annex 2 under part 96 from 28.4.2004* Bratislava

[3] Ministry of Agriculture and Rural Development 2013 *Act No. 59 from 11.3.2013* Bratislava

[4] Kobza J, Barančíková G, Bezák P, Dodok R, Grečo V, Hrivňáková K, Chlpík J, Lištjak M, Makovníková J, Mališ J, Píš V, Schlosserová J, Slávik O, Styk J, and Širáň M 2011 *Uniform analytical procedures for soil* (Bratislava: SSCRI Publishing) p. 136

[5] Wischmeier W H and Smith D D 1978 *Predicting rainfall erosion losses – Guide to conservation planning, Agricultural Handbook (1978)*, 537, (Washington: USDA Publishing) p. 58

[6] http://www.podnemapy.sk

[7] Stolbovoy V and Montanarella L 2008 Application of soil organic carbon status indicators for policy-decision making in the EU *Threats to soil quality in Europe* eds. Toth G, Montanarella L and Rusco E (Luxembourg: Joint Research Centre) JRC Scientific and Technical Reports EUR 23438 EN pp. 87-99

[8] Kobza J, Barančíková G, Čumová L, Dodok R, Hrivňáková K, Makovníková J, Náčiniaková–Bezáková Z, Pálka B, Pavlenda P, Schlosserová J, Styk J, Širáň M and Tóthová G 2009 *Soil monitoring of Slovak Republic* (Bratislava: SSCRI Publishing) p. 200

[9] Kobza J, Barančíková G, Makovníková J, Styk J, Širáň and Vojtáš J 2005 *Proposal of measures for soil protection from soil monitoring system of Slovakia results* (Bratislava: SSCRI Publishing) p.24

[10] Kobza J, Bezák P, Hrivňáková K, Medveď M and Náčiniaková Z 2007 *Criteria for identification of agricultural risk areas and methodical procedures of their evaluation* (Bratislava: SSCRI Publishing) p.40

IOP Conf. Series: Earth and Environmental Science **25** (2015) 012012 doi:10.1088/1755-1315/25/1/012012

The Conservation Effects Assessment Project (CEAP): a national scale natural resources and conservation needs assessment and decision support tool

M-V V Johnson[1], M L Norfleet[2], J D Atwood[3], K D Behrman[4], J R Kiniry[4], J G Arnold[5], M J White[5] and J Williams[6]

[1] Agronomist, United States Department of Agriculture-Natural Resources Conservation Service, 808 East Blackland Road, Temple, Texas 76502, USA

[2] Soil Scientist, United States Department of Agriculture-Natural Resources Conservation Service, 808 East Blackland Road, Temple, Texas 76502, USA

[3] Economist, United States Department of Agriculture-Natural Resources Conservation Service, 808 East Blackland Road, Temple, Texas 76502, USA

[4] Research Agronomist, United States Department of Agriculture-Agricultural Research Service, 808 East Blackland Road, Temple, Texas 76502, USA

[5] Agricultural Engineer, United States Department of Agriculture-Agricultural Research Service, 808 East Blackland Road, Temple, Texas 76502, USA

[6] Senior Research Scientist, Texas A&M Agri-Life, Blackland Research and Extension Center, 720 East Blackland Road, Temple, Texas 76502, USA

E-mail: mjohnson@brc.tamus.edu

Abstract. The Conservation Effects Assessment Project (CEAP) was initiated to quantify the impacts of agricultural conservation practices at the watershed, regional, and national scales across the United States. Representative cropland acres in all major U.S. watersheds were surveyed in 2003-2006 as part of the seminal CEAP Cropland National Assessment. Two process-based models, the Agricultural Policy Environmental eXtender (APEX) and the Soil Water Assessment Tool (SWAT), were applied to the survey data to provide a quantitative assessment of current conservation practice impacts, establish a benchmark against which future conservation trends and efforts could be measured, and identify outstanding conservation concerns. The flexibility of these models and the unprecedented amount of data on current conservation practices across the country enabled Cropland CEAP to meet its Congressional mandate of quantifying the value of current conservation practices. It also enabled scientifically grounded exploration of a variety of conservation scenarios, empowering CEAP to not only inform on past successes and additional needs, but to also provide a decision support tool to help guide future policy development and conservation practice decision making. The CEAP effort will repeat the national survey in 2015-2016, enabling CEAP to provide analyses of emergent conservation trends, outstanding needs, and potential costs and benefits of pursuing various treatment scenarios for all agricultural watersheds across the United States.
Additional Keywords: agriculture, APEX, CEAP, conservation impacts, resource assessment, simulation modelling, SWAT

1. Introduction

In the 2002 Farm Bill the United States Congress increased funding levels for conservation programs and initiatives by 80 percent compared to funding levels set in the previous Farm Bill. Administered through the United States Department of Agriculture (USDA), these funds support technical assistance and provide financial incentives to farmers and land managers to encourage voluntary conservation practice adoption on private lands. The commitment to an annual $3.5 billion dollar investment of Federal resources was accompanied by a Congressional request for greater accountability to ensure the magnitude of water quality and ecosystem service benefits achieved were commensurate with the American taxpayer investment [1]. This request exposed the need for a means of quantifying conservation practice benefits at regional and national scales. Although field-scale experiments generally supported the widespread assumption that conservation practices provide benefits, the extent to which conservation practices were achieving their intended functions of reducing edge-of-field losses of soil, nutrients, pesticides, and pathogens while supporting a sustainable and healthy agro-ecosystem was undocumented [2].

In 2003, in response to the imperative to quantify the watershed scale benefits of conservation investments, the U.S. Department of Agriculture's Natural Resources Conservation Service (USDA-NRCS) launched the Conservation Effects Assessment Project (CEAP), a multi-agency effort. CEAP's goal was to apply scientific inquiry and analyses to quantify impacts of conservation dollar investments at the national and large watershed scales. With regional and state support, CEAP continues to strengthen collaborations across academia and the interagency within five component areas: Cropland, Grazing lands, Wetlands, Wildlife, and Watersheds [3]. CEAP's Cropland National Assessment, or Cropland CEAP, focuses on quantifying impacts of cultivated cropland management and conservation practices on soil health and water quality at local, regional, and national scales. The Grazing lands component focuses on the impacts of conservation practice implementation on rangeland, pastureland, and grazed forests. The Wetlands and Wildlife components consider conservation practice impacts associated with their respective themes. The Watershed component complements all other components by providing detailed and landscape-specific assessments of a variety of benefits provided by conservation practices. These studies within the Watershed component provide scientific support for more detailed modelling of agricultural production and watershed hydrologic processes at a fine scale of resolution for selected watersheds, which contributes to improvement of regional and national scale modelling efforts [4]. Numerous academic and federal researchers are involved with CEAP related research and serve as peer reviewers for CEAP related products.

The remainder of this manuscript describes the structure and utility of the Cropland CEAP effort. Through application of a sound statistical approach to data collection and a well-validated modelling approach to estimating impacts of agricultural land management and conservation practices on natural resource concerns, Cropland CEAP has produced regional-scale reports on ten major river basins in the U.S., with reports on four additional large watersheds in development [5]. These publicly available reports provide detailed information on current (2003-2006) agricultural and conservation management and the resultant fate and transport of water, sediment, soil, nutrients, and pesticides, at the edge-of-field and landscape scales. These reports quantify the range of benefits provided by conservation practice adoption in the context of conservation concerns and ecosystem services, including water quality and soil health [6]. By informing Congress and other governmental entities of agricultural conservation successes and outstanding opportunities, these reports inform policy and improve development and implementation of future conservation programs and initiatives.

2. Data Collection for the CEAP Cropland National Assessment (2003-2006)

This section describes the data collection efforts that informed Cropland CEAP's benchmark reports. Large-scale monitoring efforts similar to some of the U.S. national effort informing CEAP are underway across the globe, and include efforts by the Australian Soil Resource Information Service (ASRIS), the Indian Bureau of Soil Survey and Land Use Planning (ICAR), the United Nation's

Office of Outer Space Affairs' Natural Resources Management and Environmental Monitoring Program (NRM&EM), and others.

The Conservation Effects Assessment Project's task was daunting in scope and complexity. In the United States in 2007 approximately 304.9 million acres were maintained in cultivated cropland and an additional 31.5 million acres of agricultural land were enrolled in the Conservation Reserve Program (CRP). Initiated in 1985, the CRP is a Federal conservation program in which enrolled farmers enter long-term contracts to remove highly erodible land from agricultural production and maintain it in continuous perennial or woody cover [7]. U.S. cultivated cropland acres support a broad variety of crops, cropping systems, and conservation practices. Field office records show over 1.6 million *new* unique combinations of conservation practices were applied to over 62.8 million cropland acres between 2003 and 2006. Estimating benefits from conservation practice adoption is further complicated by the fact that in many cases the new conservation practices were adopted on land with conservation practices already in place. Further, regional variability in soils, topography, and climate across the country contribute to variability in conservation needs and impact the magnitude of benefits derived by specific practices and suites of practices.

In the development of the Cropland CEAP benchmark reports, four data sources supported the modelling effort: the 2003 National Resource Inventory (NRI), the NRI Cropland CEAP Survey, NRCS field office records, and USDA Farm Service Agency (FAS) data [8]. CEAP was able to respond to the Congressional request for accountability so quickly because of its reliance on the successes of past monitoring efforts, including the NRI. Since 1977 the NRCS has used the NRI to provide scientific assessments of the status, conditions, and trends of natural resources across the United States in order to support agricultural and environmental policy development and program implementation. The NRI consists of a statistically designed longitudinal survey of over 200 attributes related to spatial and temporal trends in soil, water, and related natural resources on non-Federal lands in the U.S. [9]. Each NRI sampling point is linked to NRCS Soil Survey databases, which include associated soil characteristics and climate data. The well-tested, well-designed, and ever improving framework of the NRI has consistently proven to be a valuable source of nationally consistent and scientifically credible data for analysing agro-ecological impacts, economic impacts, and policy implications of land-use change [10]. Cropland CEAP's adaptation of the NRI's sampling design was logical, as it allowed CEAP to quickly establish a statistically sound national sampling protocol. A similar sampling approach could be adapted to any geographical setting.

The NRI's design supports a valid means of collecting important model inputs and a statistically sound means of assimilating and interpreting modelled results at large spatial scales [11]. A sub-sample of 31,189 NRI cropland points included in the 2002 and 2003 annual NRI surveys provided data for the national Cropland CEAP reports (Figure 1). This subsample represented about 98 percent of the diversity of cultivated cropland across the United States. However, early Cropland CEAP modelling efforts based on these NRI site specific data suggested that more information was needed to meet CEAP's goals of providing a scientifically credible assessment of agricultural conservation practice adoption impacts on the nation's natural resources [12].

To meet this need, NRCS partnered with the National Agricultural Statistics Service (NASS) to develop the NRI-CEAP Farmer Survey in order to collect more detailed information on land use and conservation practice implementation at each NRI point [13]. From 2003 to 2006 NASS-trained non-Federal enumerators administered the survey to farmers and land managers responsible for management on agricultural land at each of the 31,189 NRI cropland points. Participation in the survey was *voluntary*, with no penalties imposed on those who opted not to participate and no incentives provided to those who did participate. All proprietary information included in the survey, such as information on land management (crops grown, planting and harvesting dates, nutrient and pesticide applications, tillage applications, and conservation practice use, etc.), is held in the strictest confidence.

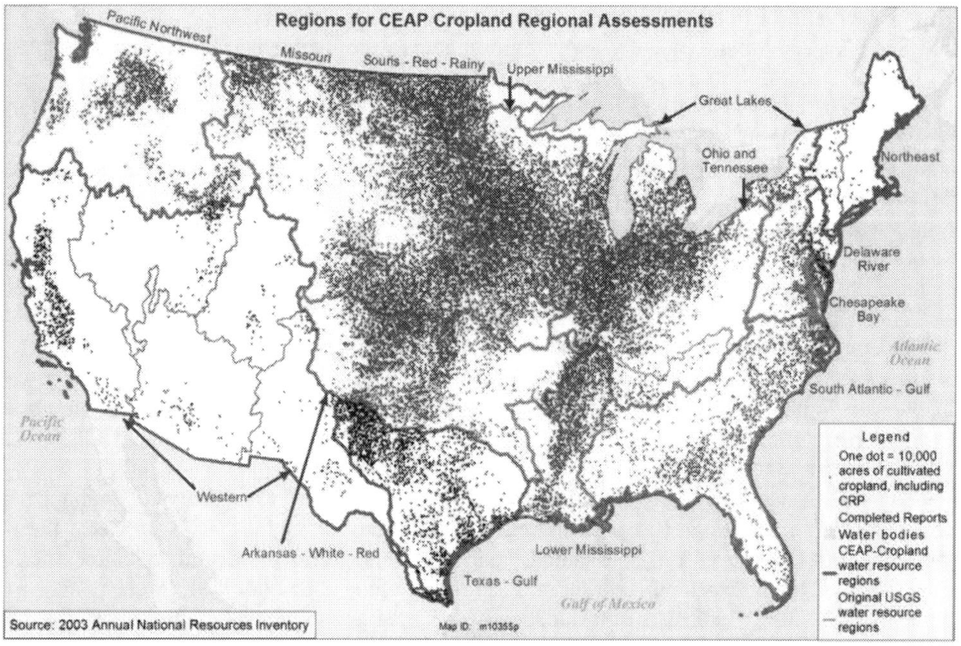

Figure 1. National Resource Inventory (NRI) points are part of a statistically sound sampling framework that allows expansion factors to be applied to data collected at each point. The NRI-CEAP points shown here are the subsample (31,189) of NRI cropland points (300,000) used to inform the benchmark CEAP reports. Cropland points include agricultural land currently enrolled in the Conservation Reserve Program (CRP), through which these acres are temporarily removed from production agriculture and maintained in perennial or woody cover. The 31,189 NRI points were sampled over the period of 2003 to 2006 and have informed all of the benchmark regional watershed scale CEAP Cropland National Assessments. Reports have been published for watersheds represented in yellow.

Across the United States NRCS field office specialists work directly with land-owners and land-managers to provide technical assistance and develop comprehensive conservation plans that are soil, climate, and land-use appropriate. These land management data, recorded by NRCS field office personnel, complemented the NRI and supplementary NASS survey data. The final source of management related model input data came from the USDA – Farm Service Agency (USDA-FSA), which provided information on agricultural lands enrolled in various conservation programs [14]. In addition to detailed land management data collected from these four sources, site specific data was required to inform the daily-time step, process based models used in CEAP analyses for simulating impacts of land management and conservation practices at each point [15].

3. The Modelling Methods in the CEAP Cropland National Assessment

The USDA-NRCS leads the Cropland CEAP modelling effort, with technical support provided by USDA-ARS and Texas A&M AgriLife. The models used in Cropland CEAP analyses for the benchmark surveys are free to the public and are readily adaptable to any geographic area. These models could be applied to CEAP-like efforts in other countries and regions.

The CEAP modelling process involves application of two daily time-step models, the Agricultural Policy Environmental eXtender (APEX) model and the Soil and Water Assessment Tool (SWAT) [16] (Figure 2). APEX and SWAT operate at different spatial scales, but interface easily. Both models operate on similar theoretical principles and are able to simulate run-off, sediment loss, leaching, plant growth, and nutrient cycling (nitrogen (N), phosphorus (P), and carbon (C) dynamics). APEX was developed to assess field-scale effects of land management and provide estimates of edge-of-field sediment, nutrient, and pesticide losses by loss pathway. APEX is a flexible model that can simulate the interactions between agricultural management and various conservation practices, including combinations of nutrient, tillage, and physical conservation management strategies. The SWAT model is a hydrologic model that considers the inputs from agricultural lands in conjunction with other land uses (*i.e.* forests, hay lands, wetlands and urban development) to assess water quality at a watershed scale. SWAT captures instream processes that occur in streams, rivers, reservoirs, and lakes, impacting the fate and transport of sediment, nutrients, and pesticides in the watershed systems [17]. Coupling APEX and SWAT provides insights into how cultivated cropland conservation practices and other land uses interact to impact the nation's waters. Both APEX and SWAT have been widely applied in studies predicting sediment, nutrient, and pollutant fate in the landscape at various spatial and temporal scales [18]. Previous work on APEX and SWAT integration provided the foundation for integration of the models used in Cropland CEAP's national effort [19].

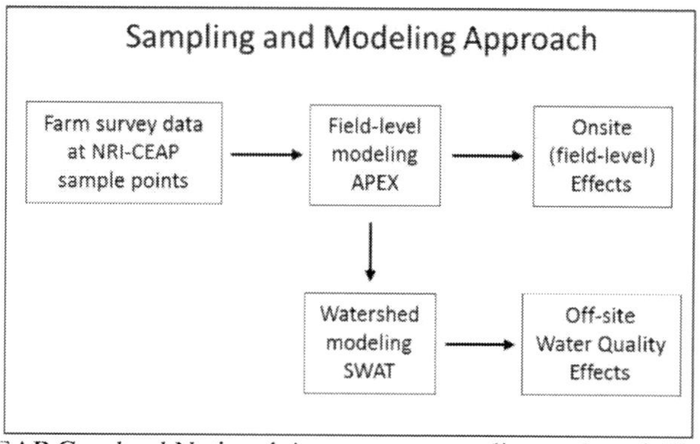

Figure 2. The CEAP Cropland National Assessment sampling and modelling framework.

In the Cropland CEAP modelling framework, the APEX model is used to estimate the impacts of reported land management and conservation practice adoption on factors that impact soil health and water quality, including soil water holding capacity, C sequestration and water and nutrient retention on farmed fields. The majority of scientific research on conservation practice effects has been conducted at the field scale upon which APEX operates, enabling modellers to develop well calibrated and validated simulations. The watershed component of CEAP continues to provide scientific data to improve field scale modelling efforts [20]. APEX simulates structural conservation practices, including in-field overland flow controls, in-field concentrated flow controls, and edge-of-field filter and buffering controls. APEX also simulates residue and tillage management, including conservation tillage, mulch tillage, and conventional tillage. N and P management methods are also simulated, including simulation of rate of application, method of application, nutrient source, and timing of application. APEX accommodates changes in crop rotations, including cover crop and winter cover adoption. Specifically, APEX simulates the interactions between these management approaches and conservation practices to predict plant growth and water, nutrient, and pesticide dynamics to estimate yield stability and water, N, P, C, sediment, and pesticide losses at the edge-of-field, including below the root-zone [21].

SWAT uses aggregated APEX output as input representing cultivated cropland contributions to instream loads of runoff, nutrients, pesticides, and sediments. SWAT is able to accommodate other land use inputs as well, including point and non-point source discharge to simulate the interaction between all land use types in the upstream watersheds. SWAT then routes the cumulative flows and loads through each downstream watershed and eventually routes the water, sediment, and nutrients through the river basin outlet. As the water, nutrients, and sediments move through the system, SWAT applies in-stream and reservoir routing submodels to simulate instream load dynamics and delivery from one watershed to the next. This enables modellers to estimate the impacts of cultivated cropland conservation practice adoption on sediment and nutrient fate and transport across various regional scales [22]. Appropriate delivery ratios are applied to account for the fact that nutrient and sediment deposition may contribute to legacy loads, which may later become re-suspended or dissolved back into the water column [23]. Instream processes simulated by SWAT include sediment degradation, streambed deposition, nutrient cycling and transformation, and other in-stream kinetic processes [24]. Details on APEX and SWAT modelling processes, assumptions, validation, and calibration techniques applied within Cropland CEAP simulations are available in the literature [25].

By using APEX outputs to inform the SWAT simulations, Cropland CEAP is able to answer larger scale policy relevant questions, including assessment of conservation practice impacts at the watershed scale. The APEX outputs may be aggregated and scaled up to the eight-digit watershed scale with the same expansion factors developed for expansion of other NRI data [26]. The eight-digit hydrologic watersheds (HUCs) delineated by the United States Geological Service (USGS) are the scale at which current Cropland CEAP sampling and modelling protocol produce statistically sound reports. From a statistical perspective, performing the weighted aggregation of data after the APEX model runs is preferable to data aggregation prior to model simulation, as the latter practice may mask the actual statistical distribution of data [27].

4. The Benchmark CEAP Cropland National Assessment Reports

Cropland CEAP has generated a series of regional reports that provide a baseline estimate of conservation practice adoption and resources status across the United States, based on analyses of the 2003-2006 NASS NRI survey and associated data [28]. The scientifically defensible design of the NRI, upon which the CEAP survey was based, allows the APEX and SWAT models to simulate the anticipated impacts that long-term adoption of current (2003-2006) conservation practices at a regional scale. Process based models were setup to simulate reported land use, land management, and conservation practices at a national scale, in order to determine the edge-of-field and watershed impacts of agricultural production on water quality, water quantity, erosion rates, and nutrient and carbon dynamics. Current conservation treatments and outstanding treatment needs could be compared across the nation's cropland to help inform current and future policy and practice implementation. Cropland CEAP developed a classification system of treatment needs for cropland acres based on inherent site vulnerability and the current level of conservation treatment in place (Figure 3). "Adequately treated" or "low needs" acres had annual average per acre loss rates of less than 2 tons of sediment per year, 15 pounds of N through runoff losses, 25 pounds of N through leaching, and 3 pounds of P to runoff. When model simulations showed that an acre's level of vulnerability exceeded the current conservation effort on that field, causing loss rates to exceed those just listed, the acre was classified as having outstanding conservation treatment needs, with high needs acres showing the highest levels of imbalance between inherent vulnerability and current conservation. A discussion of the development of this classification structure is beyond the scope of this manuscript, which intends to demonstrate the benefits to conservation planning that may be derived from a CEAP or CEAP-like assessment of cultivated cropland within any geographical setting. Further information on the system of classification used to determine conservation need levels is available in the regional CEAP reports [29].

The CEAP classification system was applied to cultivated cropland represented by each NRI-CEAP point. Analyses of the acreage included in each of the nine published reports demonstrate great variability in conservation treatment needs across the nation's river basins (Figures 1 and 3). On average, 54 percent of the cropland acres in these nine basins were adequately treated as of 2003-2006, but acres with outstanding conservation treatment need were not uniformly distributed across the landscape. For example, per this classification system, over 80 percent of acres in the Missouri Basin were adequately treated, but fewer than 14 percent of the acres in the Lower Mississippi Basin were adequately treated. Because the Basins were selected based on drainage rather than size, they varied in size from 83.6 million acres (Missouri Basin) to 4.3 million acres (Chesapeake Bay Region). Although within the Upper Mississippi Basin only 15 percent of the acres have high treatment needs, because of the large size of the basin, these acres account for 27 percent of the United States' high needs acres (Figure 4).

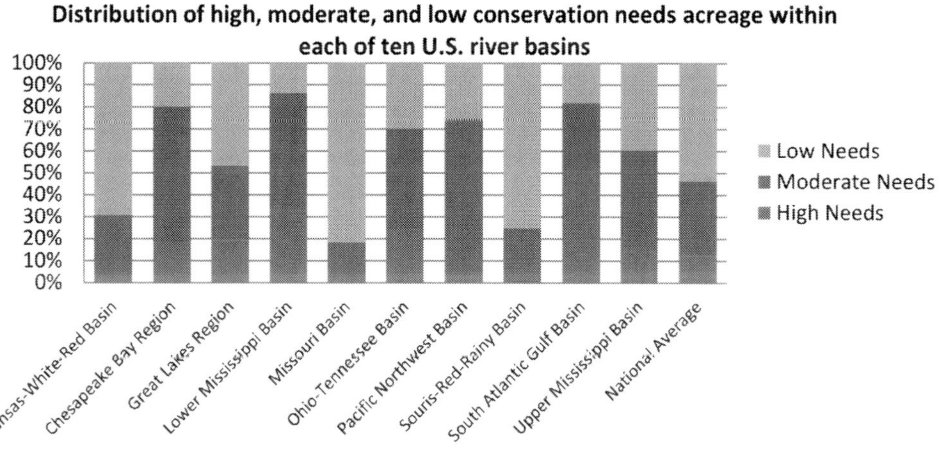

Figure 3. The percentage of cultivated cropland acreage in each basin identified as having High, Moderate, or Low need for additional conservation treatment to address one or more conservation concern based on inherent site vulnerability and current conservation practices.

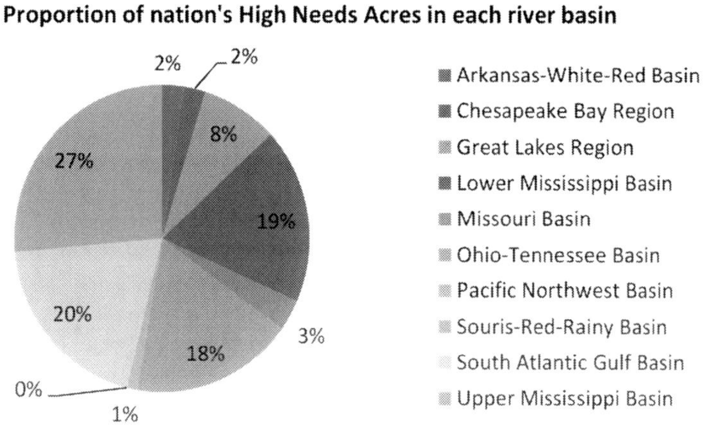

Figure 4. The percentage of the United States' cultivated cropland acreage classified in the CEAP classification system as having high outstanding needs for additional conservation practices, as distributed in each of ten river basins analyzed in CEAP's Cropland National Assessment.

In the seminal CEAP Cropland National Assessment regional reports the Cropland CEAP national survey provided information for only one point in time, the 2003-2006 sample period. These data provide an important baseline against which future conservation may be compared. However, in order to provide Congress with an estimation of current conservation practice benefits, the CEAP effort constructed a hypothetical "no practice" scenario in which all reported agricultural conservation practices were removed and replaced with reasonable conventional management. Choices in management in the no-practice scenario were made with care to guard against overestimation of current practice benefits. For example, a reported no-till system would be simulated as two tandem disking operations 3 days prior to spring planting as opposed to intense fall tillage with a mouldboard plow. More information on construction of the no-practice scenario may be found in the regional reports [30]. All other model inputs and parameters were maintained between the two scenarios, allowing the modellers to attribute the differences between the 2003-2006 baseline condition simulation and the no practice condition simulation to cultivated cropland conservation practices [31]. The APEX model enabled per-acre quantification of conservation practice impacts on surface water run-off rates and percolation volumes, water and wind erosion rates, nutrient dynamics and loss pathways, and carbon dynamics. The SWAT model interpreted cultivated cropland conservation practice impacts on these same dynamics at the watershed scale, providing insights into the role of agricultural conservation impacts in context with other land use impacts at the watershed scale. The APEX and SWAT models are extremely flexible and can accommodate additional alternative management scenarios. In fact, various CEAP reports explored simulations of targeting all high needs or all high and moderate needs acres (per the CEAP needs classification system previously discussed) [32].

5. Beyond the CEAP's Original Intent: a Decision Support Tool Emerges

The regional Cropland CEAP reports address the original driver of CEAP development: they assess the benefits derived by conservation practices in place during the 2003-2006 survey. However, the iterative CEAP process has also improved the modelling capacities of APEX and SWAT to act as powerful decision support tools within the CEAP context [33]. The versatility of these models and their ability to provide information on impacts of changing conservation practices enables the identification of conservation gains and remaining conservation needs at multiple spatial scales. Within a given spatial scale, the models may be used to explore anticipated impacts of various schemes of prioritizing conservation treatment.

Budget disbursement decisions are complicated and take a number of factors into consideration, including political and economic realities. The simplest means of allocating funds, equally among all basins is likely not the most efficient for achieving conservation gains. Each basin varies in terms of size, current level of treatment, and outstanding conservation needs. If funds were disbursed on the basis of acreage in cropland, the Upper Mississippi and Missouri would be allocated 21 and 30 percent of all funding (Figure 5). A more informed means of prioritizing acreage for treatment could include selecting to treat acreage with the most significant outstanding needs first. Here we used the CEAP needs classification system discussed above, which was developed solely for use in CEAP reporting, to enable comparison across basins. In this scheme the Upper Mississippi basin's allocation would increase to 27 percent of the budget and the Missouri basin's allocation would decline to only 3 percent (Figure 4). Simulated outcomes for treating all high needs acreage or all high and moderate needs acreage are included in each of the regional reports [34].

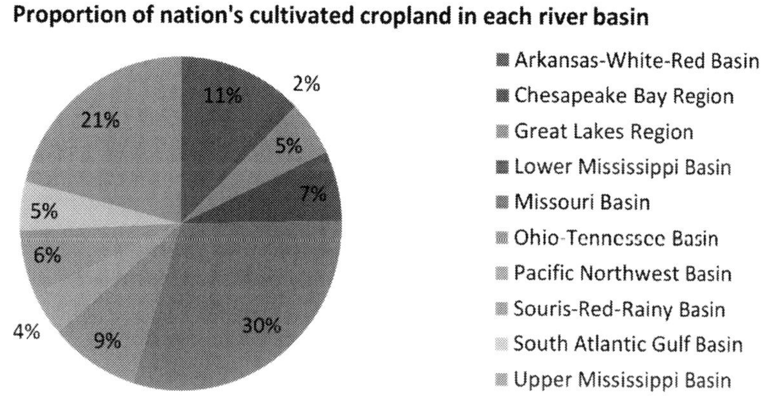

Figure 5. The percentage of the United States' cultivated cropland acreage, as distributed in each of ten river basins analyzed in in CEAP's Cropland National Assessment.

Another means of selecting acreage for treatment is choosing to address particular conservation concerns. Conservation concerns vary by region across the United States, with some areas more prone to water or wind driven erosional losses, while others are more prone to leaching losses. This means some areas tend to lose soil and nutrients through runoff loss pathways (water, N, P, C, and sediment), while others have more significant problems with leaching losses (water and N). Conservation programming in the United States has traditionally focused on controlling erosion, but in the 1990's the focus began to shift to be more inclusive of all natural resources related to the agroecosystem. By coupling the APEX model with mathematical programming to enable optimization of conservation treatment application, Cropland CEAP outputs allow modelers to explore the potential impacts of shifting the focus of conservation practice adoption across a broad spectrum of conservation concerns, including combinations of two or more conservation concerns (Figure 6). Here we demonstrate that N loss reductions could be maximized subject to also achieving other environmental benefits, such as sequestration of C; the results of this analyses suggest focusing on the dual conservation concerns of N loss reduction and C sequestration would provide 98 percent of the N loss reduction benefits and more than double the C sequestration benefits as would focusing solely on N loss reduction (Figure 6). Using a multi-criteria optimization model allows maximization of multiple interacting conservation goals. An optimization approach is flexible and could be designed to accommodate a variety of restrictions and concerns, including limiting acreage selection, limiting budget disbursements, focusing on particular loss pathways, selecting for particular conservation treatments, etc. The optimization model could be increased in complexity to allow weighting of various conservation goals. Determining the best set of weights for various conservation goals is beyond the scope of this analysis.

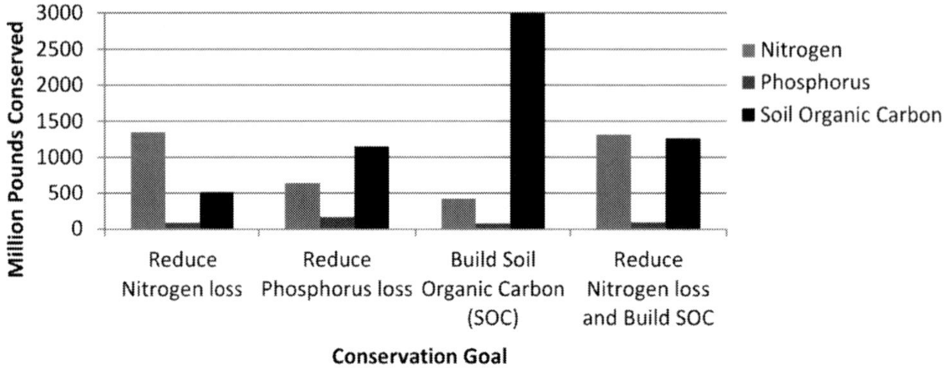

Figure 6. Potential national level impacts of shifting conservation goals between various conservation concerns, assuming a fixed budget and full participation by land owners on acres selected for treatment. Each conservation approach provides ancillary benefits to conservation concerns other than the primary concern on which the conservation goal is focused.

The ability of the models to represent watersheds within the national context also enables exploration of conservation needs as they vary geographically. In this modelling exercise no controls were placed on acreage or funding distributions across basins. Because different parts of the country have different conservation needs, prioritizing one conservation concern over another at the national scale skews funding and acres treated to the parts of the country most severely in need of increased conservation efforts to address that particular concern (Figure 7). This figure demonstrates the fact that different regions of the country have different conservation concerns and changing conservation priorities will change fund distribution patterns. Policy makers considering efforts to address nutrient and soil losses are necessarily concerned with both the conservation outcomes and the distribution of funds in a geopolitical context.

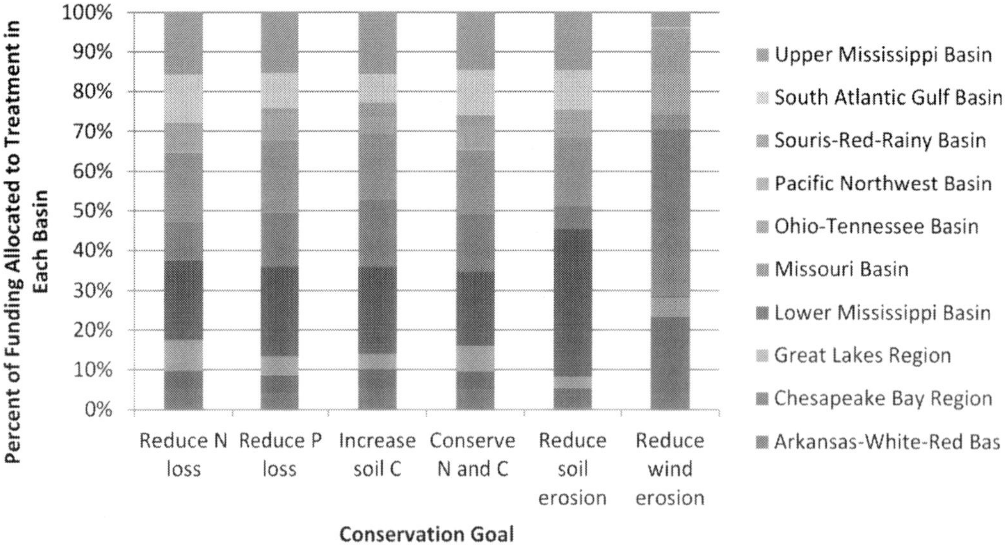

Figure 7. Shifting priorities from one conservation concern to another shifts funding and treated acreage from one basin to another, as different parts of the country have different severities of need to address each conservation concern. This figure demonstrates how conservation treatment would shift

between basins depending on whether the emphasis was on nitrogen loss reduction, phosphorus loss reduction, building soil organic carbon, adopting the dual focus of nitrogen loss reduction in conjunction with building soil organic carbon, reducing soil erosion, or reducing wind erosion.

6. Conclusion

The benchmark Cropland CEAP reports provide a baseline against which all impacts of all current and future conservation practices and programming may be assessed. Although the farm survey data used to generate the Cropland CEAP assessments is not publicly available, the benchmark Cropland CEAP assessments are all available online [35]. These reports continue to contribute to development of improved decision support tools, refined sampling protocols, revisions to national databases, and modelling improvements. A recent resurvey of the Chesapeake Bay watershed emphasizes the tremendous value of the quantified benchmark reports, against which conservation successes may be measured [36].

In addition to addressing the Congressional request for quantification of conservation practice impacts, Cropland CEAP has explored myriad hypothetical conservation practice scenarios, all of which would have cascading impacts on funding disbursements, geographical distribution of conservation practices and programming, and selection of suites of appropriate conservation practices to meet conservation goals. It is not the purpose of this manuscript to select the best conservation treatments for U.S. cultivated cropland acreage, nor to choose the best conservation goal, or means of selecting that goal. Here we demonstrate how these models and data may be used to construct and explore conservation scenarios that provide land managers and policy makers with information they need to make decisions about future conservation efforts at multiple scales in order to meet the concerns of the farmer, the local community, the environment, the state, and the American taxpayer.

Cropland CEAP provides a powerful decision support tool for United States policy makers, enabling efficient application of the American taxpayer's conservation dollar. The resurvey of the country slated to occur in 2015 and 2016 will further improve this powerful tool. Reanalysis of the benchmark data (2003-2006) alongside the resurvey will improve our scientific understanding of the impacts of and trends in conservation practices at the local and watershed scales. Further, Cropland CEAP's second survey and analyses of the nation's state of conservation will increase the scientific knowledge base helping policy makers implement appropriate programs and helping land managers and farmers apply appropriate practices to best meet conservation goals. Thus Cropland CEAP has evolved from its original goal of assigning metrics to past conservation efforts to become a dynamic and promising tool informing current and future conservation decision making in the United States.

Acknowledgements

The United States Department of Agriculture (USDA) is an equal opportunity provider and employer. Mention of trade names or commercial products in this publication does not imply recommendation or endorsement by the USDA.

References

[1] Tomer M D and Locke M A 2011 *Water Sci Technol* **64** 300-10

[2] Mausbach M J and Dedrick A R 2004 *J Soil Water Conserv* **59** 96A-103A

[3] Maresch W, Walbridge M R and Kugler D 2008 *J Soil Water Conserv* **63** 198A-203A

[4] Richardson C W, Bucks D A and Sadler E J 2008 *J Soil Water Conserv* **63** 590-604

[5] United States Department of Agriculture, National Resources Conservation Service, Conservation Effects Assessment Project, Cropland National Assessment webpage (http://www.nrcs.usda.gov/wps/portal/nrcs/detail/national/technical/nra/ceap/?cid=nrcs143_014144)

[6] United States Department of Agriculture, Natural Resources Conservation Service, Soil and Water Resources Conservation Act Report 2011 (United States Department of Agriculture:

Washington, DC) p 164 (http://agwaterconservation.colostate.edu/PDFs/2011_RCA_Appraisal_Pre_Publication_Copy.pdf)

[7] United States Department of Agriculture, Natural Resources Conservation Service, National Resources Inventory 2009 *Summary Report: 2007 National resources inventory,* (Washington, DC and Center for Survey Statistics and Methodology, Iowa State University, Ames, Iowa) p 123 (http://www.nrcs.usda.gov/Internet/FSE_DOCUMENTS/stelprdb1041379.pdf)

[8] United States Department of Agriculture, Natural Resources Conservation Service, National Resources Inventory webpage (http://www.nrcs.usda.gov/wps/portal/nrcs/main/national/technical/nra/nri/results/)

[9] Nusser S M and Goebel J J 1997 *Environ Ecol Stat* **4** 181-204

[10] Goebel J J and Kellogg R L 2002 Using survey data and modelling to assist the development of agri-environmental policy *Conf. on Agricultural and Environmental Statistical Applications in Rome* (Rome, Italy: National Statistical Institute of Italy) pp 695-704

[11] Goebel J J 2009 *Statistical methodology for the NRI-CEAP Cropland Survey* p 28 (http://www.nrcs.usda.gov/Internet/FSE_DOCUMENTS/nrcs143_013402.pdf)

[12] Potter S, Andrews S, Atwood J D, Kellogg R L, Lemunyon J, Norfleet L and Oman D 2006 *Model simulation of soil loss, nutrient loss, and change in soil organic carbon associated with crop production* United States Department of Agriculture, Natural Resources Conservation Service, Washington, DC p 245 (http://babel.hathitrust.org/cgi/pt?id=umn.31951d027280886;view=1up;seq=1)

[13] United States Department of Agriculture, Natural Resources Conservation Service, 2006 CEAP Farmer Surveys. (United States Department of Agriculture, Natural Resources Conservation Service: Washington, DC) (http://www.nrcs.usda.gov/wps/portal/nrcs/detail/national/technical/nra/ceap/?cid=nrcs143_014163)

[14] Allen A W and M W Vandever, eds 2005 *The Conservation Reserve Program - Planting for the future* (Reston, Virginia: United States Geological Survey, Biological Resources Discipline, Scientific Investigations Report) p 268 (http://www.fsa.usda.gov/Internet/FSA_File/21490.pdf)

[15] Gianessi L P and Peskin H M 1984 *An overview of the RFF environmental inventory: methods, sources, and preliminary results v. 1* (Renewable Resources Division, Resources for the Future: Washington, DC) p 111

[16] Arnold J G, Srinivasan R, Muttiah R S and Williams J R 1998 *J Am Water Resour As* **34** 73-89

[17] White M J, *et al.* 2014 *J Soil Water Conserv* **69** 26-40

[18] Gassman P W, Reyes M R, Green C H and Arnold J G 2007 *T ASABE* **50** 1211-50

[19] Saleh A and Gallego O 2007 *T ASABE* **50** 1177-87

[20] Duriancik L F, *et al.* 2008 *J Soil Water Conserv* **63** 185A-197A

[21] Wang X, Kannan N, Santhi C, Potter S R, Williams J R and Arnold J G 2011 *T ASABE* **54** 1281-98

[22] Arnold J G, Williams J R and Maidment D R 1995 *J Hydraul Eng-ASCE* **121** 171-183

[23] Wang X, Kannan N, Santhi C, Potter S R, Williams J R and Arnold J G 2011 *T ASABE* **54** 1281-98

[24] Neitsch S L, Arnold J G, Kiniry J R, Williams J R and King K W 2002 *Soil and Water Assessment Toll (Version 2000). Theoretical documentation* GSWRL 02-01, BRC 02-05, TR 191. (Texas Water Resources Institute: College Station, Texas)

[25] Santhi C, Kannan M, White M, DiLuzio M, Arnold J G, Wang X and Williams J R 2014 *J Environ Qual* ***43*** 177-98

[26] Goebel J J 2009 *Statistical methodology for the NRI-CEAP Cropland Survey* pp 28 (http://www.nrcs.usda.gov/Internet/FSE_DOCUMENTS/nrcs143_013402.pdf)

[27] Goebel J J and Kellogg R L 2002 Using survey data and modelling to assist the development of agri-environmental policy *Conf. on Agricultural and Environmental Statistical Applications in Rome* (Rome, Italy: National Statistical Institute of Italy) pp 695-704
[28] United States Department of Agriculture, National Resources Conservation Service, Conservation Effects Assessment Project, Cropland National Assessment webpage (http://www.nrcs.usda.gov/wps/portal/nrcs/detail/national/technical/nra/ceap/?cid=nrcs143_014144)
[29] United States Department of Agriculture, National Resources Conservation Service, Conservation Effects Assessment Project, Cropland National Assessment webpage (http://www.nrcs.usda.gov/wps/portal/nrcs/detail/national/technical/nra/ceap/?cid=nrcs143_014144)
[30] United States Department of Agriculture, National Resources Conservation Service, Conservation Effects Assessment Project, Cropland National Assessment webpage (http://www.nrcs.usda.gov/wps/portal/nrcs/detail/national/technical/nra/ceap/?cid=nrcs143_014144)
[31] Duriancik L F, *et al.* 2008 *J Soil Water* Conserv **63** 185A-197A
[32] United States Department of Agriculture, National Resources Conservation Service, Conservation Effects Assessment Project, Cropland National Assessment webpage (http://www.nrcs.usda.gov/wps/portal/nrcs/detail/national/technical/nra/ceap/?cid=nrcs143_014144)
[33] Arnold J G, Harmel R D, Johnson M V V, Bingner R, Strickland T C, Walbridge M, Santhi C, DiLuzio M and Wang X 2014 *J Soil Water Conserv* **69** 137A-44A
[34] United States Department of Agriculture, National Resources Conservation Service, Conservation Effects Assessment Project, Cropland National Assessment webpage (http://www.nrcs.usda.gov/wps/portal/nrcs/detail/national/technical/nra/ceap/?cid=nrcs143_014144)
[35] United States Department of Agriculture, National Resources Conservation Service, Conservation Effects Assessment Project, Cropland National Assessment webpage (http://www.nrcs.usda.gov/wps/portal/nrcs/detail/national/technical/nra/ceap/?cid=nrcs143_014144)
[36] United States Department of Agriculture, Natural Resources Conservation Service 2013 Impacts of Conservation Adoption on Cultivated Cropland in the Chesapeake Bay Region, 2003-06 to 2011 Pp 113 (http://www.nrcs.usda.gov/wps/portal/nrcs/detail/national/technical/nra/ceap/?cid=stelprdb1240074)

Using satellite imagery to asses trends in soil and crop productivity across landscapes

K Sheffield[1] and E Morse-McNabb[2]

[1]Agriculture Research Division, Department of Economic Development, Jobs, Transport and Resources, Parkville, Victoria, Australia
[2]Agriculture Research Division, Department of Economic Development, Jobs, Transport and Resources, Bendigo, Victoria, Australia

E-mail: kathryn.sheffield@ecodev.vic.gov.au

Abstract. Measuring different indicators of production and soil health over the long term will help build a picture of soil health and productivity across the landscape. This paper examines the potential contribution of satellite imagery to this area. This investigation undertook a very long time series analysis of Landsat imagery (approximately 40 years) and MODIS imagery (approximately 10 years). Novel datasets and approaches were used to assess areas based on land use history and land cover condition. Spring Normalised Difference Vegetation Index (NDVI), land cover maps based on NDVI thresholds, annual cumulative NDVI and fractional ground cover (FGC) were used to identify trends in vegetation cover change at a landscape scale, and their relationship with factors such as land use intensification history, geomorphology, rainfall, and land use. This work has improved the broad, baseline understanding of production variation across the landscape, while also providing a practical demonstration of the integration of a range of disparate data sources.

1. Introduction

Farming systems and land use changes, including production intensity and land management, can result in soil changes that compromise current and future capacity for primary production and provision of ecosystem services. Climate and land management affect net primary production (agricultural productivity), that in turn influences soil processes and health. Agricultural production impacts many soil properties such as soil structure, bulk density, soil microbes and soil carbon [1-4]. It is important to ensure soil health is not negatively impacted by agricultural practices to sustain productivity in the long term. Converting land use from native vegetation to agriculture production results in multiple changes, including increases in soil carbon and nutrients from crop residue and fertiliser inputs, and changes in soil microbial activity, plant production and other aspects of soil quality [4, 5].

Measuring different indicators of production and soil health over the long term will help build a picture of soil health and productivity across the landscape. This paper examines the potential contribution of satellite imagery to this area. Benefits of remotely sensed data include continual spatial coverage across the landscape, enabling variation within, and between, paddocks to be identified; long archives of data, allowing spatial trends over time to be monitored; readily available data acquired at frequent time intervals; and the ability to integrate information with other sources of spatial data such as soil and landform mapping, and climatic data. There are many different spatial data sets available that can contribute to an understanding of land use and production history in Victoria, including soil and landform information, climatic data such as rainfall, long-term production history [6] and remotely sensed data. Investigations into the relationships between these types of data have improved understanding of the linkages between remotely sensed data and landscape and agricultural information, such as crop rotation, productivity, soil properties and land management practices [7-9].

Satellite imagery, which records information in the visible and near-infrared portions of the electromagnetic spectrum, captures conditions at the earth's surface, such as vegetation cover. Variables derived from satellite imagery infer information about soil condition by observing what happens on the surface in terms of vegetation growth. Differences in vegetation cover linked to agricultural productivity in farming landscapes can be attributed to a number of factors, including climatic conditions, soil properties, geomorphological factors, that occur as gradations across the landscape, and also land cover and management, occurring as sharp transitions aligning with land parcels [8-10]. Information derived from remotely sensed data, such as the Normalised Difference Vegetation Index (NDVI), has been used as an indicator of green vegetation cover [8, 11-13]. Total NDVI accumulation within a year, Spring vegetation cover (NDVI) and fractional ground cover (FGC) are examples of measures that can be derived from remotely sensed data, and that may provide useful information about agricultural productivity and changes in soil health. FGC is also an important factor in wind and water erosion mitigation [4].

This study aims to improve our understanding of the influence of land cover history and various environmental factors on soil condition in the Wimmera region of north-west Victoria, by examining linkages between vegetation cover over time (derived from satellite imagery) and factors such as rainfall and production history. Identifying factors that affect agricultural production can highlight areas where production falls short of expectations, and provide insights into impacts from soil health and land management factors. An improved understanding of land cover history will increase our understanding of soil condition, and linkages between soil health and the productive capacity of Victorian agricultural landscapes.

2. Methods

2.1. Study area

The study area is located in the Wimmera and Southern Mallee region of Victoria (Figure 1). The study area is positioned in an 'overlap' area of the Landsat satellite acquisition path, meaning that images are potentially acquired at twice the frequency compared with areas not located in an overlap area. Data used to generate the annual cumulative NDVI product, rainfall data, and land privatisation information were available over a larger spatial extent and clipped to a wider area of interest (Figure 1).

The study area is dominated by rain-fed dryland cropping production systems. Approximately 97% of the study area is privately owned, being used largely for cropping (approximately 90%) and grazing. There are smaller areas of irrigation agriculture and timber production. A small proportion of the study area remains as Crown Land, including parks, reserves and riparian vegetation consisting of remnant native vegetation.

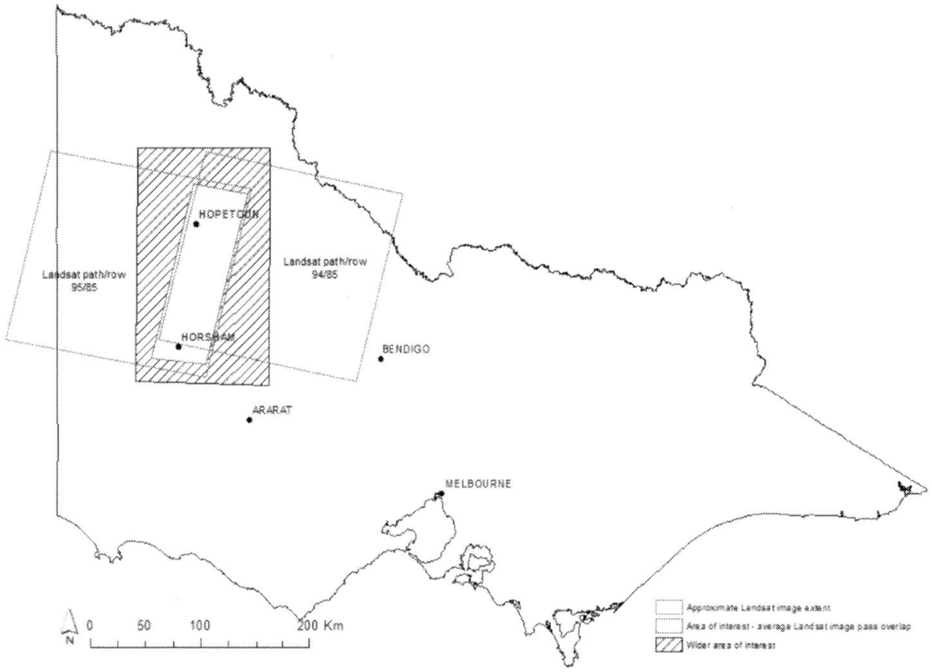

Figure 1. Study area in the Wimmera and Southern Mallee region of Victoria, Australia showing approximate study area and image extents.

2.2. *Data*

This study used 97 Landsat images acquired between 1973 and 2012, sourced from the United States Geological Survey (USGS) Landsat archive. Landsat images have a swath (approximately 170 km x 180 km) large enough to deliver landscape scale information, and a spatial resolution (approximately 30 m – 60 m) which is able to deliver information at a paddock and sub-paddock scale. The 'overlap' area of the Landsat satellite acquisition path between two Landsat scenes (path/row: 94/85 and 95/85) was used for the study area as this increased the potential frequency of image acquisitions. The duration of the image series (i.e. over 40 years) enables longer-term trends of vegetation growth and land use change to be detected. Imagery captured by Landsat sensors 1-7 were used, with each image calibrated to top of atmosphere reflectance using published coefficients [14]. Current advancements in processing of the Landsat image archive have led to a reliably calibrated data set, with a consistent absolute radiometric scale between Landsat sensors and total uncertainties of under 10% for most sensors and bands [15, 16]. Further geo-rectification was undertaken as required and all images resampled to a 30 m spatial resolution to provide a consistent spatial unit across the time series. Clouds and shadow were masked from each image and NDVI calculated using red and near-infrared (NIR) reflectance using equation (1):

$$NDVI = \frac{NIR - Red}{NIR + Red}$$

Where: NIR = near-infrared reflectance
Red = red reflectance (1)

Each year was considered in terms of seasons: Summer (December to February), Autumn (March to May), Winter (June to August) and Spring (September to November) and, where available, one cloud-free image per season was acquired. This process generated a series of 97 NDVI images spanning from 1973 to 2012. Over this 40 year time period, 29 images were acquired during Summer, 20 in Autumn, 18 during Winter, and 30 in Spring. This study examined Spring vegetation cover, with dense green vegetation considered to have an NDVI value greater than 0.65. This threshold was estimated based on observations of local information and published NDVI crop growth curves [11, 12, 17]. The time series of 30 Spring images was grouped into five-year time intervals, and the percentage of time a pixel was covered by green vegetation (NDVI>0.65) was calculated, producing a measure of high NDVI frequency.

A Fractional Ground Cover (FGC) product was generated using the Joint Remote Sensing Research Program algorithm, hosted on the AusCover web site [18] that estimates bare ground (BG), photosynthetic vegetation (PV) and non-photosynthetic vegetation (NPV) from Landsat imagery across different seasons. This study used Autumn seasonal median NPV product.

In addition to information derived from Landsat imagery, MODIS imagery was also utilized. The MOD13Q1 16-day vegetation product (NDVI) was downloaded from the USGS archive for 2003-2012. A single MODIS image tile provides complete spatial coverage of the state of Victoria. The MODIS NDVI images were clipped to Victoria and re-projected, and the data was then summed within a calendar year to create an annual cumulative NDVI product for 2003-2012. This product is used as a surrogate measure of biomass accumulation. The MODIS cumulative NDVI product provides a summation of all NDVI values within a year, which is distinct from the Landsat NDVI product which is based on NDVI for a single image date. Therefore the two NDVI-based products are measured on different scales, both of which are unitless.

Other spatial data used included land privatization (i.e. when land tenure initially changed from Crown Land to privately owned land), and land use in 2005, estimated from Landsat images acquired between 1989 and 2005 [6]. Land privatization was considered according to three time periods: before 1866 (early pastoralism), between 1866 and 1888, and after 1888 (Victorian gold rush and agricultural expansion) [6]. Detailed of the development of the data set and sources of data used to compile this information are published in [6]. Annual rainfall data was obtained from the Bureau of Meteorology and was interpolated from point data to form a 5 km gridded data set. Rainfall classes used in this study were: <300 mm, 300-350 mm, 350-400 mm and >400 mm.

3. Results and discussion

Factors that affect the soil environment in agricultural settings include land management (cultivation and post-harvest stubble practices), agricultural inputs (fertilisers, chemicals, and animal waste if the crop is grazed), crop or pasture type, and climate variables such as rainfall and temperature. Agricultural production impacts many soil properties such as soil structure, bulk density, soil carbon and soil nutrients, which influence plant production and soil quality [1-5]. Long term productivity is also linked to the soil microbial population, which is influenced by crop type and land management options such as stubble retention [5].

Measuring different indicators of production and soil health over the long term builds a picture of soil health and productivity across the landscape. Annual accumulated NDVI and Spring vegetation cover are examples of measures able to be derived from remotely sensed data which may provide useful information about agricultural productivity and changes in soil health. In this study, crop frequency is used as an indicator of production intensity, while annual accumulation of NDVI is used as an indicator of production biomass.

Figure 2 illustrates fluctuations in vegetation cover over time, and differences between land uses. Dryland agriculture and grazing show similar patterns with small fluctuations over time, while irrigated agriculture shows more variability in high NDVI frequency, most likely related to water availability. Timber production shows less frequent fluctuations, related to harvesting and planting cycles within the plantations. Vegetation located on Crown Land, which is predominantly native woody scrub vegetation, had a low average percentage of time with a Spring NDVI<0.65 and very little fluctuation between time periods (Figure 2), suggesting that vegetation cover on Crown Land, while typically lower than agricultural land, has remained relatively steady over the past four decades. This also illustrates how agriculture has changed vegetation cover trends within the landscape.

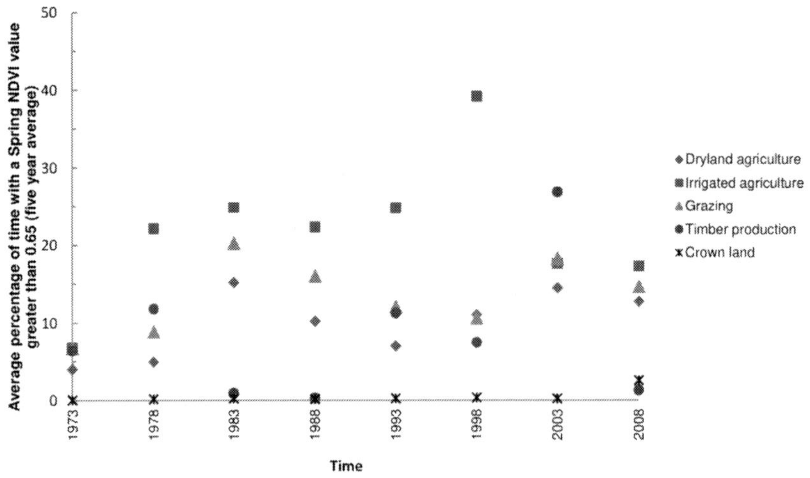

Figure 2. Average percentage of time with a Spring NDVI value >0.65 (five-year average), from 1973 to 2012, based on land use in 2005 (land use classes from [6])

The remaining results relate to dryland agriculture land use only. Figure 3 compares annual accumulated rainfall classes with the percentage of time the Spring NDVI is >0.65 in a 5-year time period. The trend is variable, but six of the eight time periods show that areas that recorded the highest amount of rainfall also had the highest percentage of time with dense Spring vegetation. The association between cumulative NDVI and rainfall class varies between years but generally higher rainfall results in higher annual cumulative NDVI (Figure 4). Variation between rainfall classes is inconsistent, with smaller variation in some years (2005) compared with others (2008 and 2012). This variation may be influenced by the change in spatial location of the rainfall classes over time. There is a general gradient of lower rainfall in the north to higher rainfall in the south, but the spatial distribution of the different rainfall classes changes annually. The data shows that not every year has all rainfall classes, such as in 2010 when the entire study area had an annual accumulated rainfall in excess of 400 mm. Therefore, the rainfall classes incorporate a different mix of soil types and land management histories in each year which will contribute to the variation seen in the data. There is a strong link between accumulated NDVI and rainfall >400 mm in all years except 2006.

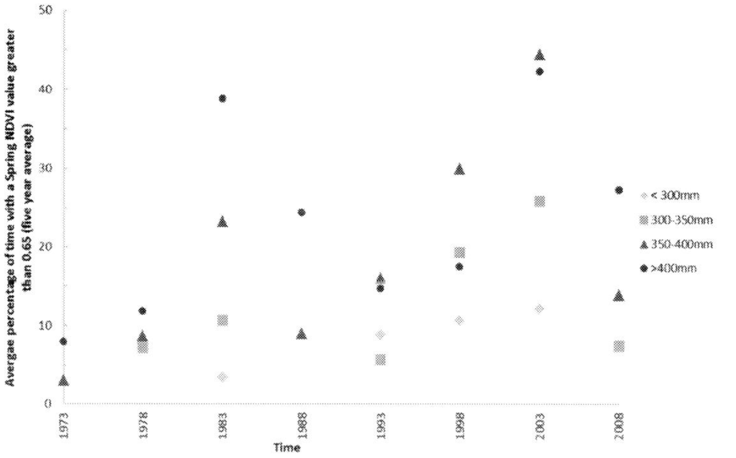

Figure 3. Average percentage of time with a Spring NDVI value >0.65 (five-year average), from 1973 to 2012, based on rainfall classes (dryland agricultural land)

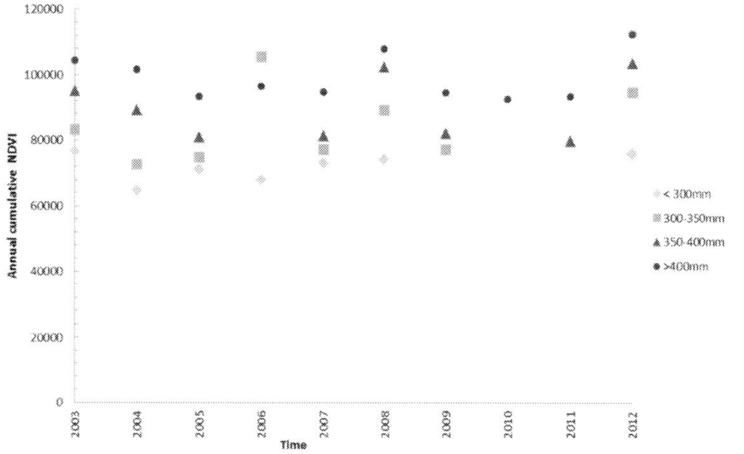

Figure 4. Annual cumulative NDVI (calculated from MODIS imagery) for dryland agriculture from 2003 to 2012 based on annual rainfall classes

Land use history, including production intensity and land management, can result in soil changes that compromise current and future capacity for primary production, provision of ecosystem services, and soil properties such as carbon content [2-4]. A challenge in documenting these impacts is that they occur over extended periods of time. Differences in agricultural vegetation cover can be attributed to a number of factors, including climatic conditions and soil properties, that occur as gradations across the landscape, and also land cover and management, which are often seen as sharp transitions aligning with land parcels [8-10].

Trends in vegetation cover were also analysed with respect to land privatisation. Both Spring vegetation cover and annual accumulated NDVI show a clear stable response to privatisation date throughout the whole time series. These analyses identify some factors that affect agricultural production and highlight areas where production falls short of expectations. Land put into agricultural production earlier had a higher percentage of time with dense Spring vegetation in a five-year time period compared with land that was privatised later (Figure 5). Six out of the eight time periods in the

40 year history showed this trend (Figure 5). This response is influenced by the relationship between land privatisation and rainfall gradient and a potentially higher frequency of crops.

There is a strong consistent relationship between land privatisation period and annual cumulative NDVI, with areas privatised from Crown Land before 1866 always showing higher cumulative NDVI than areas privatised later (Figure 6). This is similar to trends shown in the 40 year Landsat data, although differences appear to be more consistent in the MODIS imagery. This may be due to the improved temporal resolution of the MODIS data. Land which falls into the first privatisation date class, i.e. prior to 1866, covers less than 2% of the study area, which reduces the variation in this class compared with the other two land privatisation periods (1866-1888 and after 1888). However, the relationship between privatisation date and vegetation production holds for the second and third periods. The link between increased vegetation cover and privatisation date seems to be mostly due to rainfall gradient, as rainfall is a dominant factor in vegetation growth. However, in 2010 the entire study area received a similar amount of rainfall (Figure 4), but there are clear differences in biomass accumulation when land privatisation date is considered (Figure 6). This demonstrates that other factors such as soil health and land management may also influence these trends. Understanding that there are inherent limitations to production in the landscape suggests that strategies to improve soil condition and productivity need to be recommended on a regional basis, and land use history can provide a simple and effective method of identifying properties which may face similar challenges.

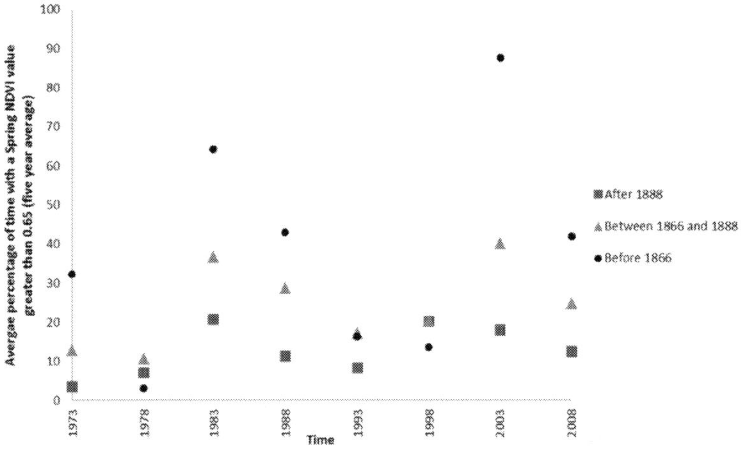

Figure 5. Average percentage of time with a Spring NDVI value >0.65 (five-year average), from 1973 to 2012, based on time privatised from Crown Land holdings (dryland agricultural land)

Land management practices such as post-harvest cover retention, combined with crop rotation management, have been shown to increase production biomass and soil microbial activity, and assist in long term maintenance of soil health [19]. NPV ground cover in Autumn is used as an indicator of ground cover retention post-harvest as soil surface cover is a key factor for a number of soil physical properties, stabilises soil, provides protection from wind and water erosion, nutrient retention, organic matter content and carbon fixation [1, 2, 4]. NPV remaining after harvest is also a source of nutrients for soil microorganisms [5]. Less crop biomass reduces the quality and quantity of post-harvest stubble, which is a major source of soil organic carbon [4]. Stubble retention can also indicate particular land management practices, such as minimum tillage.

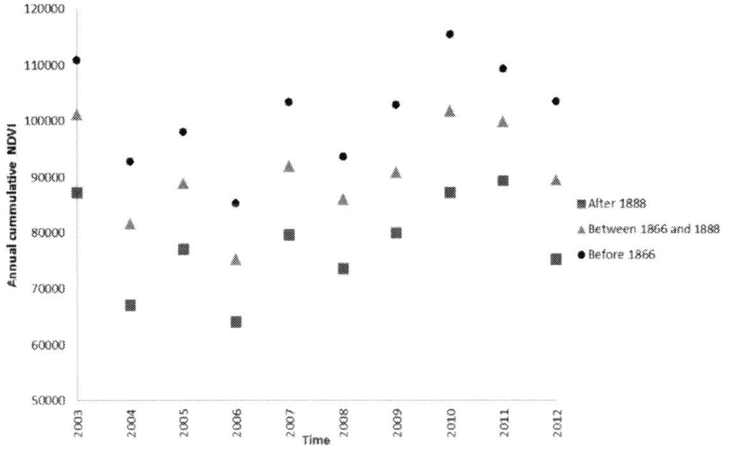

Figure 6. Annual cumulative NDVI (calculated from MODIS imagery) for dryland agriculture between 2003 and 2012 based on time privatised from Crown Land holdings

Trends in NPV ground cover and biomass accumulation showed that higher biomass accumulation in the preceding year led to higher levels of NPV ground cover in the following Autumn. To further explore this variation, and establish the effects of ground cover retention on crop productivity in subsequent years, further analysis of the season, crop type and biomass accumulation is necessary. Areas with higher NDVI accumulation from the previous year (i.e. >80,000) showed consistently higher NPV cover in the following Autumn compared to areas with lower biomass accumulation from the previous year (Figure 7). The pattern is similar to that shown in accumulated biomass from the following year (Figure 8), however further investigation of climatic factors and land management practices may explain some observed discrepancies in this relationship (i.e. 2003 and 2011).

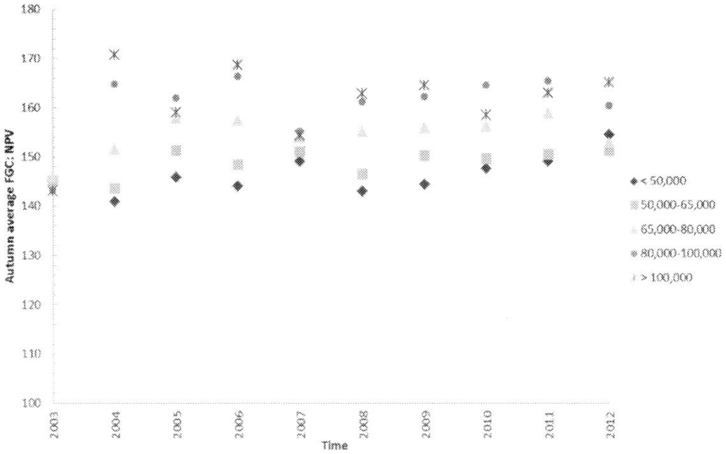

Figure 7. Autumn average FGC: NPV for dryland agriculture from 2003 to 2012, based on annual cumulative NDVI derived from MODIS imagery from the previous year.

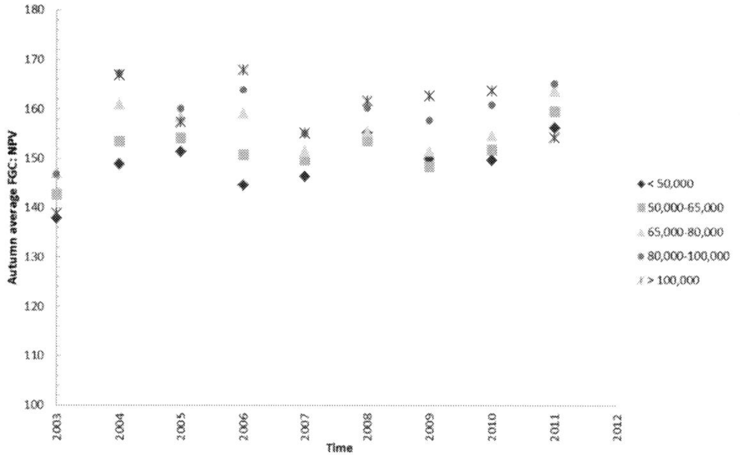

Figure 8. Autumn average FGC: NPV for dryland agriculture from 2003 to 2012, based on annual cumulative NDVI derived from MODIS imagery from the following year

4. Conclusion

Measuring different indicators of production and soil health over the long term will help build a picture of soil health and productivity across the landscape. This paper examined the potential contribution of satellite imagery to this area by examining trends in total NDVI accumulation within a year, Spring vegetation cover and FGC over an extended period of time (i.e. 10-40 years). Benefits of remotely sensed data include continual spatial coverage across the landscape, enabling variation within and between paddocks to be identified; long archives of data, allowing spatial trends over time to be monitored; readily available data acquired at frequent time intervals; and the ability to integrate information with other sources of spatial data. This study aimed to improve our understanding of soil health in the Wimmera region of north-west Victoria, by examining linkages between vegetation cover over time and factors such as rainfall and production history.

Vegetation growth was strongly linked with annual rainfall, with higher rainfall generally corresponding with higher annual cumulative NDVI and more frequent dense Spring vegetation growth. However, variations within this general pattern suggests other factors contribute to vegetation growth. Both Spring vegetation cover and annual accumulated NDVI show a clear stable response to land privatisation date throughout the time series, suggesting that land privatisation has an influence on production and therefore soils, either as a legacy of the base characteristics of the land or as a consequence of long-term management of the land. Trends in NPV ground cover and biomass accumulation showed that higher biomass accumulation in the preceding year led to higher levels of NPV ground cover in the following Autumn, which has important consequences for both soil health, erosion mitigation, and agriculture production. Identifying factors that affect agricultural production can highlight areas where production falls short of expectations, and provide insights into soil health and land management factors. An improved understanding of land cover history will increase our understanding of soil condition, and linkages between soil health and the productive capacity of Victorian agricultural landscapes.

References

[1] Allen DE, Singh BP, Dalal RC 2011 Soil Health Indicators Under Climate Change: A Review of Current Knowledge. In: Sing BP, et al, editors. Soil Health and Climate Change. Berlin, Heidelberg: Springer-Verlag;.

[2] Kibblewhite MG, Ritz K, Swift MJ 2008 Soil health in agricultural systems. *Phil Trans R Soc B*. **363**: 685-701

[3] Kasel S, Bennett LT 2007 Land-use history, forest conversion, and soil organic carbon in pine plantations and native forests of south eastern Australia. *Geoderma*. **137** 401-13
[4] Cork S, Eadie L, Mele P, Price R, Yule D 2012 The relationships between land management practices and soil condition and the quality of ecosystem services delivered from agricultural land in Australia. Canberra, Australia: Kiri-ganai Research Pty Ltd.
[5] Gupta VVSR, Neate SM, Leonard E 2007 Life in the Soil. Glen Osmond, South Australia: Cooperative Research Centre for Soil & Land Management, incorporating the CSIRO, The University of Adelaide, the South Australian Minister for Primary Industries (SARDI and PISA) and the State of Victoria (Department of Natural Resources & Environment: Agriculture Victoria).
[6] Sinclair SJ, White MD, Medley J, Smith E, Newell GR 2012 Mapping the past: Constructing a digital land-use history map for Victoria, Australia. *Proc Roy Soc Vic* **124** (3) 193-206
[7] Dang YP, Pringle MJ, Schmidt M, Dalal RC, Apan A 2011 Identifying the spatial variability of soil constraints using multi-year remote sensing. *Field Crops Res* **123** 248-58
[8] Hill MJ, Donald GE 2003 Estimating spatio-temporal patterns of agricultural productivity in fragmented landscapes using AVHRR NDVI time series. *Remote Sensing of Environment*. **84** 367-84
[9] Sumfleth K, Duttmann R 2008 Prediction of soil property distribution in paddy soil landscapes using terrain data and satellite information as indicators. *Ecol Indic* **8** 485-501
[10] Wen L, Yang X, Saintilan N 2012 Local climate determines the NDVI-based primary productivity and flooding creates heterogeneity in semi-arid floodplain ecosystem. *Ecol Model* (242) 116-26
[11] Clark ML, Aide TM, Grau HR, Riner G 2010 A scalable approach to mapping annual land cover at 250 m using MODIS time series data: A case study in the Dry Chaco ecoregion of South America. *Remote Sens Environ* **114** (11) 2816-32
[12] Wardlow BD, Egbert SL, Kastens JH 2007 Analysis of time-series MODIS 250m vegetation index data for crop classification in the U.S. Central Great Plains. *Remote Sens Environ* **108** 290-310
[13] Cihlar J, St-Laurent L, Dyer JA 1991 Relation Between the Normalized Difference Vegetation Index and Ecological Variables. *Remote Sens Environ* **35**: 279-98
[14] Chander G, Markham BL, Helder DL 2009 Summary of current radiometric calibration coefficients for Landsat MSS, TM, ETM+, and E0-1 ALI sensors. *Remote Sens Environ* **113** 893-903
[15] Hansen M, Loveland TR 2012 A review of large area monitoring of land cover change using Landsat data. *Remote Sens Environ* **122** 66-74
[16] Markham BL, Helder DL 2012 Forty-year calibrated record of earth-reflected radiance from Landsat: A review. *Remote Sens Environ* **122** 30-40
[17] United States Geological Survey. Remote Sensing Phenology 2011 [23/9/2014]: Available from: http://phenology.cr.usgs.gov/ndvi_foundation.php.
[18] Joint Remote Sensing Program. Fractional ground cover for Australia derived from USGS Landsat images. 2014 [14/4/2014]; Available from: http://data.auscover.org.au/xwiki/bin/view/Product+pages/Landsat+Fractional+Cover.
[19] Govaerts B, Mezzalama M, Unno Y, Sayre KD, Luna-Guido M, Vanherck K, et al 2007 Influence of tillage, residue management, and crop rotation on soil microbial biomass and catabolic diversity. *Appl Soil Ecol* **37** 18-30

Soil moisture monitoring for crop management

Dale Boyd
Department of Economic Development, Jobs, Transport and Resources, PO Box 441, Echuca 3564, Australia

E-mail: Dale.Boyd@ecodev.vic.gov.au

Abstract. The 'Risk management through soil moisture monitoring' project has demonstrated the capability of current technology to remotely monitor and communicate real time soil moisture data. The project investigated whether capacitance probes would assist making informed pre- and in-crop decisions. Crop potential and cropping inputs are increasingly being subject to greater instability and uncertainty due to seasonal variability. In a targeted survey of those who received regular correspondence from the Department of Primary Industries it was found that i) 50% of the audience found the information generated relevant for them and less than 10% indicted with was not relevant; ii) 85% have improved their knowledge/ability to assess soil moisture compared to prior to the project, with the most used indicator of soil moisture still being rain fall records; and iii) 100% have indicated they will continue to use some form of the technology to monitor soil moisture levels in the future. It is hoped that continued access to this information will assist informed input decisions. This will minimise inputs in low decile years with a low soil moisture base and maximise yield potential in more favourable conditions based on soil moisture and positive seasonal forecasts

1. Introduction

There are limited examples of the use of soil moisture probes in a dryland cropping system in Victoria but increasing interest in the grains sector from both service providers and farmer groups.
A scoping study conducted by the Department of Primary Industries (DEPI) on the focus groups of farmers involved in the projects nine monitoring sites and found that:

Soil moisture levels are estimated/measured by over 90% of the group and conducted by a number of methods, generally subjective. Most popular were estimations from recent rainfall events, crop condition, and drive by observations. Some were using self calibrated tools such as push probes and soil sampling and determining the wetness of soil. General feeling was that estimations could be improved on current methods and enough cases where they had got it wrong to explore alternatives.

Agronomists generally had a greater interest in soil moisture monitoring technology as a need to measure, when advising on critical business decisions. Benefits identified were with time efficient monitoring and the ability to compare previous year's data. There were a very small percentage of farmers who were using volumetric sampling where millimetres of water could be calculated with some farmers using these measurements in modelling programs such as yield prophet.

A state-wide project across many rainfall districts and soil types, there are some differences identified by the groups of farmers with use of probes. In low rainfall zones, using soil moisture probes will aid the crop choice decisions by showing pre sowing plant available. Good soil moisture reserves also provides confidence to sow by the calendar. Higher rainfall zones with reliable winter rainfall, farmers will follow rotations as guided by Best Management Practices and not by soil

IOP Conf. Series: Earth and Environmental Science 25 (2015) 012014 doi:10.1088/1755-1315/25/1/012014

moisture reserves but knowing the soil moisture levels will allow strategic inputs through the growing season to target the potential high yields.

2. Material and Methods

Growers current cropping systems may not be maximising water use efficiency, if they are using subjective assessments. The practice of using moisture probes has had limited use in the grains industry and as such many farmers are unaware to utilise this technology. Moisture sensors and telemetry devices that use the mobile network to send the data to be securely stored on a server that is accessed with the internet. Farmer focus groups were generated based on recognized cropping regions and linked to local sites to assist in the validation of examining deep soil moisture and determining usefulness, usability, and availability. Participants were educated in interpretation of data, received regular email updates explaining recent soil water changes and had access to live data via a soil moisture graphing website.

3. Data Access

Access to this data enabled project participants -

- Observe deep soil moisture (30-100cm) at one representative point in a common soil type on a district farm
- Moisture displayed in absolute soil moisture content and with the iNTELLiGraph software, crop upper and lower limits under different soil types and crop types could be estimated.

4. Results

Over the three seasons of monitoring the project has built up knowledge of estimated crop upper and lower limits in different soil types across different crops. The seasons were quite extreme with at least one wet summer (some sites with two and floods). The majority of the growing seasons have been low decile and has seen a huge depletion of soil moisture reserves through late August to October. The value of sub-soil moisture has been clearly evident and participants have been amazed of the ability of crops to use moisture from 60cm and beyond and how quickly a large biomass crop will use moisture in spring. The following are examples of where this information could be used to assist in decision making in dryland cropping.

Site at Youanmite (North East Victoria), dotted red oval indicates the depletion of soil moisture reserves through Spring 2011 and the dashed blue oval represents the flooding rains in early March 2012 that refilled the profile. The star indicates the sowing time where the crop was planted with a good profile of deep moisture. The green shaded area is where plant moisture use has been observed, and the white area below is nearing crop lower limit and moisture stress. The black line is a summed total of soil water content from 30-100cm.

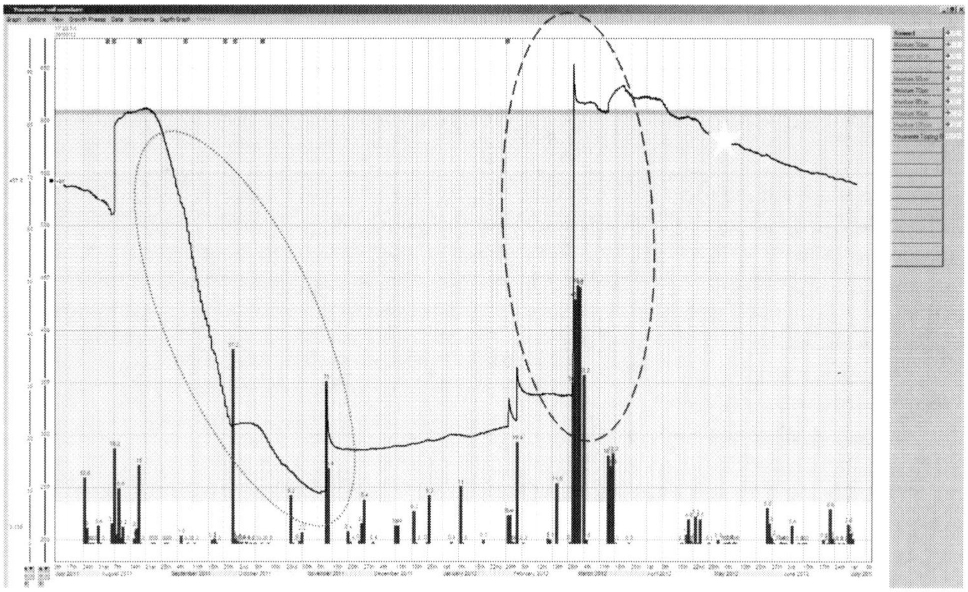

Figure 1. Accumulation of soil moisture through summer and autumn from a dry soil profile.

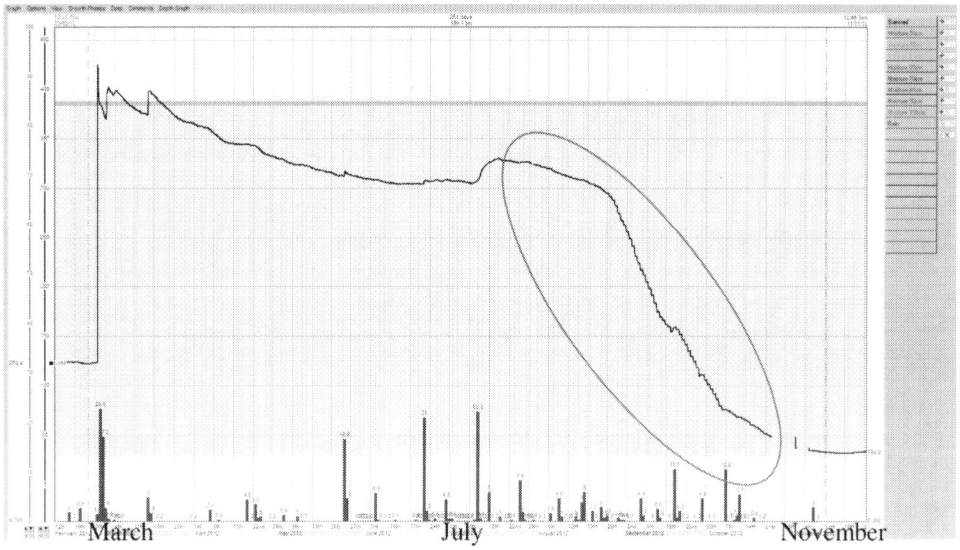

Figure 2. Summed soil moisture graph displaying plant water use through spring with high levels of plant available water.

Site at Raywood (Central Victoria), the red oval highlights the depletion of a near full profile of moisture recorded in July 2012 with a low decile spring. The level recorded in November was an estimated crop lower limit for that monitored soil type of clay loam.

IOP Conf. Series: Earth and Environmental Science **25** (2015) 012014 doi:10.1088/1755-1315/25/1/012014

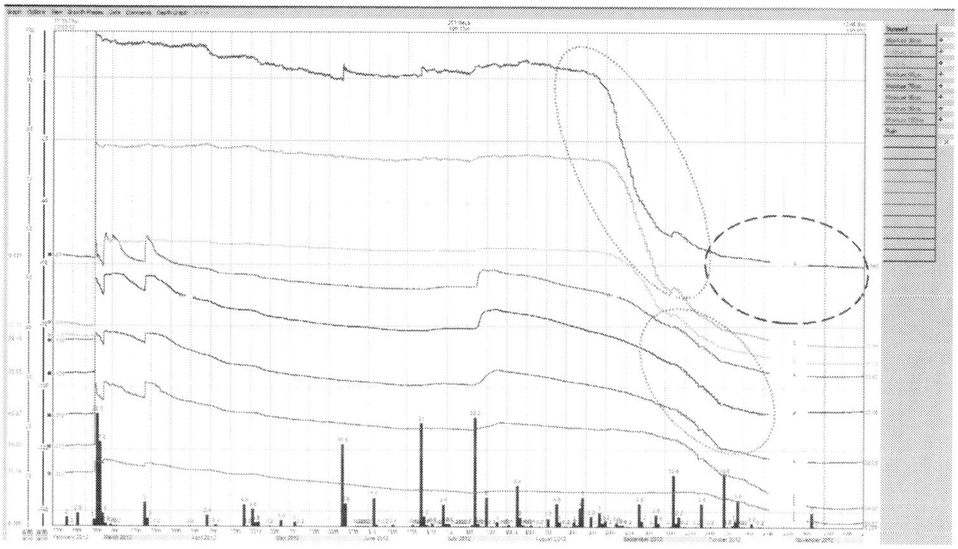

Figure 3. Individual sensor stacked graph of the Raywood SMM site.

A zoomed view of the Raywood site of all the individual sensor depths (positioned every 10cm starting at 30cm to 100cm). Dotted ovals highlight the moisture use from the different soil depths. Early spring moisture is used from the easiest area of access being the shallow sensors while in late September moisture is being used from 60cm down to a meter. The dashed oval indicates that the sensors at that depth has reached crop lower limit with no change in the soil water content (flat line) during October. At this site with the dry spring, crop lower limit was reached down to one meter. (NB. separate sensor graphs are defined by colour and sensor lines are arranged as per depth location, shallowest on top and deepest on the bottom).

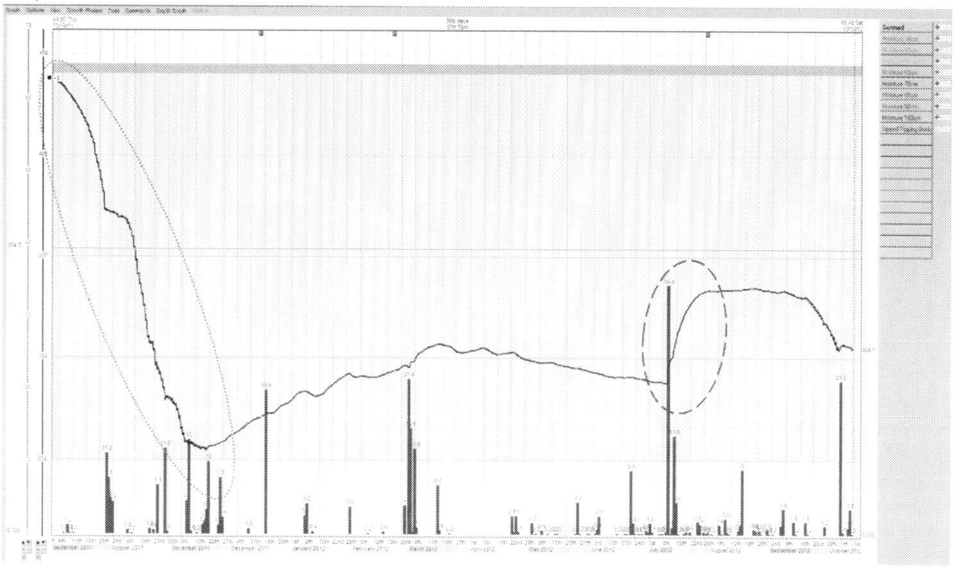

Figure 4. Varied water use through two spring periods associated with different levels of plant available water.

Summed soil moisture content from site at Speed (Mallee), dotted red oval highlights the depletion of a full profile of moisture in 2011 with a low decile spring. The soil moisture reserves were never built up to those levels again even after significant rain in July deposited moisture deep (dashed blue oval). Low moisture use in spring 2012 was attributed to crop vigour and limited plant available moisture.

5. Conclusions

The challenge to the grains sector is to increase productivity growth. This may be from a number of options, some of which include:

- Better farm decisions through improved information - not just more information
- Potentially through the uptake of new technology.

The soil moisture monitoring network has proven to successfully link to both of these with positive results. The challenge is Micro monitoring a site for paddock size management. Different soil types across paddock and farm and how many monitoring sites required versus the ability to vary paddock management.

What can legacy datasets tell us about soil quality trends? Soil acidity in Victoria

B P Marchant[1], **D M Crawford**[2] **and N J Robinson**[3]
[1] British Geological Survey, Nicker Hill, Keyworth Nottingham, UK
[2] Department of Economic Development, Jobs, Transport and Resources, Ellinbank, Victoria, Australia
[3] Department of Economic Development, Jobs, Transport and Resources, Bendigo, Victoria, Australia

E-mail: benmarch@bgs.ac.uk

Abstract. Purpose-built soil monitoring networks have been established in many countries to identify where soil functionality is threatened and to target remediation initiatives. An alternative to purpose-built soil monitoring networks is to use legacy soils information. Such information yields almost instant assessments of soil change but the results should be interpreted with caution since the information was not collected with monitoring in mind. We assess the threat of soil acidification in Victoria using two legacy datasets: (i) the Victorian Soils Information System (VSIS) which is a repository of the results of soil analyses conducted for scientific purposes since the 1950s and (ii) a database of 75 000 routine soil test results requested by farmers between 1973 and 1993. We find that the VSIS measurements are clustered in space and time and are therefore suitable for local rather than broad-scale assessments of soil change. The farmers' results have better spatial and temporal coverage and space-time models can be used to quantify the spatial and temporal trends in the pH measurements. However, careful validation of these findings is required since we do not completely understand how the measured paddocks were selected and we cannot be certain that sampling or laboratory protocols have not changed with time.

1. Introduction

Around the world there are threats to soils and the important functions which they perform. These include loss of organic matter, erosion, sealing, compaction, loss of biodiversity, contamination, landslides and salinization. There is concern that these threats are increasing in intensity and number, perhaps as a result of climate change. The soil must be monitored to determine when and where these threats are emerging and remediation is required. Much of this monitoring is initiated locally by farmers and other land managers who use the resultant information to decide what measures are required to maintain the productivity of their soils. There is also an important role for broader-scale soil monitoring networks to determine more general trends. Such information permits governments to devise targeted policies to maintain soil quality where it is under threat and to test whether such policy measures have been effective. Governments might provide education, resources and other support to land owners so that they can effectively protect their soils or introduce legislation to prevent practices which harm the wider environment.

For these reasons many countries have introduced national-scale soil monitoring networks (SMNs). These are designed to ensure that unbiased estimates can be made of both the status and change of soil properties. There are costs associated with the implementation of a SMN although they are small in comparison to productivity loses and environmental damage that can result from soil degradation. Changes in soil properties are often slow and it is generally necessary to wait a decade or more before resampling the networks and testing whether relatively stable properties such as soil organic carbon content have undergone change that is large enough to be reliably detected. For this reason few SMNs have currently completed their second phase of sampling. One exception is the National Soil Inventory of England and Wales [1] which has been used to quantify the decreases in soil acidity that have resulted from decreased acid deposition.

Policy makers are aware of the long-term need for purpose-built SMNs (e.g. [2]) but they require immediate information about the rate of soil change and given that many soil changes are irreversible, policy-makers cannot afford to wait more than a decade until newly implemented SMNs have been resampled. Therefore many countries are exploring what they can learn about soil change from legacy soil information (e.g. [3]). This might involve the resampling of soil inventory sites that were used for past mapping exercises or the interrogation of other databases of soil information such as the results of soil analyses requested by farmers. Some caution is required when using this legacy information. In general the data were not collected with monitoring in mind. Soils exhibiting certain characteristics could have been preferentially sampled and these preferences might vary in time. For example a farmer might focus his soil sampling efforts where he thinks nutrient inputs are required and he might change where he samples within his farm as he learns more about the management of soils and fields on his farm. Therefore, there is potential for naïve analyses of such information to lead to biased results.

We illustrate these points with reference to the problem of soil acidification in Victoria, Australia. In 2004 the Environment and Natural Resources Committee of the Parliament of Victoria conducted an inquiry into the impacts and trends in soil acidity [4]. The inquiry estimated that soil acidification was leading to an annual loss of AU$ 470 million of agricultural production in Victoria. At that time approximately 23 % of Victoria's agriculturally productive soils were thought to be affected by acidity and it was forecast that this area could double over the next 20 years. Therefore a number of initiatives were implemented to reverse the trend. These included a campaign to raise awareness of the problems of acid soils, attempts to reduce the costs of soil tests in regions most affected by acid soils, efforts to ensure consistency between soil analysis laboratories and research into how intensive agricultural practices can impact on soil pH. The estimates and forecasts contained in the report were uncertain. The enquiry highlighted that a scarcity of data hampered their efforts to quantify the threat of acidification. Therefore, as part of their Understanding Soil and Farming Systems project (http://vro.depi.vic.gov.au/dpi/vro/vrosite.nsf/pages/lwm_usfs), the Victorian Department of Environment and Primary Industries (DEPI; formerly DPI) explored the potential to use legacy information to quantify soil acidification. In this paper we discuss two sources of legacy information. The first is the Victorian Soils Information System (VSIS), a repository of soil pit data collected since the 1950s by Victorian Government soil surveyors. The second is a dataset of 75,000 soil analyses requested by farmers in Victoria between 1973 and 1993 [5]. We explore the extent to which these data might be used to map the status and change of topsoil pH across Victoria.

2. Background

2.1. Soil Monitoring Networks

Arrouays et al. [6] describe how many countries (e.g. France, UK, Denmark, Austria, Switzerland and Germany) have adopted national-scale purpose-built SMNs. There are two distinct types of design for these purpose-built networks. In *probabilistic designs* the sampling locations are selected randomly and independently of each other whereas the locations for a *purposive design* are selected to best serve a specific objective. Often a purposive design consists of a regular grid since this ensures that the samples are evenly dispersed over the study area and are hence suitable for producing maps.

The main advantage of probabilistic designs is that they can be analyzed by classical statistical methods which require very few assumptions and lead to unbiased estimates of the mean and variance of the status or rate of change of property of interest. If these classical methods are applied to data collected according to a purposive design then biased estimates can occur. For example, if a soil property is sampled on a regular grid then the sample variance is likely to be an inflated estimate of the actual variance of the property within the study region. This is because the grid design does not include any points that are located less than the grid-spacing apart. Hence for a purposive design, it is necessary to estimate a statistical model of the spatial correlation of the property so that such a bias can be removed. Any inferences or conclusions drawn about the variation of the soil property are only reliable if the statistical model is an appropriate representation of the soil property. Therefore thorough validation of the statistical model is required. Note that in the case of a regular grid the average of the collected data is an unbiased estimate of the mean of the soil property because locations are not preferentially sampled according to any attribute of the soil.

Arrouays et al. [6] listed examples of legacy data being used to monitor soil change. These involved either the resampling of inventory sites or the use of soils information that was collected for another purpose (e.g. measurements to assist farmers make fertiliser management decisions). However, the original purposes for collecting legacy data are not necessarily the same as purposes for which SMN are established. For instance, farmers might focus their soil analyses where they expect there to be a deficit of nutrients. Such preferential sampling could lead to a bias in classical statistical estimates of both the mean and the variance of the property of interest. One might hope to remove such biases through the use of a suitable statistical model but this is only possible if the criteria used to select the locations of samples are well understood. Often this information is not recorded.

2.2. The threat of soil acidity and soil acidification

Soil pH is a measure of the acidity or alkalinity of the soil. It is either measured in water or a weak solution of calcium chloride using one part soil to five parts aqueous solution. Measurements made in calcium chloride are less affected by seasonal variations. They can be 0.6-0.8 units lower than measurements made in water although this difference can be less in saline soils. In this paper we generally quote calcium chloride values unless we state otherwise.

The ideal soil pH for plant growth is between 5.5 and 6.5. Outside this range the pH influences availability of elements such as phosphorus, molybdenum, zinc, aluminum and manganese and can lead to toxic levels or deficiencies. Soil acidity can adversely affect plant production and limit a farmer's choices when sowing crops or pastures to acid tolerant plant species. Soil acidity can also have a wider effect on the environment. Reduced vegetative cover can lead to increased leaching, runoff and erosion, with concomitant off-site effects such as increased ascensions to groundwater and increased movement of nitrates leading to contamination of groundwater and surface water. Soils tend to be more acidic in high rainfall regions due to increases leaching of nitrates and other ions.

When undesirably low soil pH is measured, the acidity can be ameliorated by applying lime. Soil acidification can be moderated through balancing inputs and outputs of organic matter such as hay, the selection of crop and pasture species and careful fertilizer management to improve nitrogen cycling. These measures are mainly implemented by land managers at the farm- or paddock-scale but policy makers can facilitate them through the provision to farmers of incentives such as subsidized soil testing, and education.

2.3. The Victorian Soils Information System (VSIS)

The Victorian DEPI has recently collated soil pit data collected for various projects and research purposes into a repository known as VSIS. This database is currently available to DEPI staff and in the future will be made available to natural resource managers, scientists and modellers outside DEPI. The location of each soil pit has been recorded. The VSIS contains 1415 observations of topsoil pH that are comparable with the farmers data described below. These samples extend from the surface to between 8 and 12 cm and were extracted between 1958 and 2011. Figure 1 (left) shows the temporal and spatial coverage of these samples. Samples from each time period tend to be clustered in space reflecting the local focus of the different research projects (such as the Victorian pasture portion of the National Soil Fertility Project, NSFP, shown in figure 1 right) which contributed to VSIS.

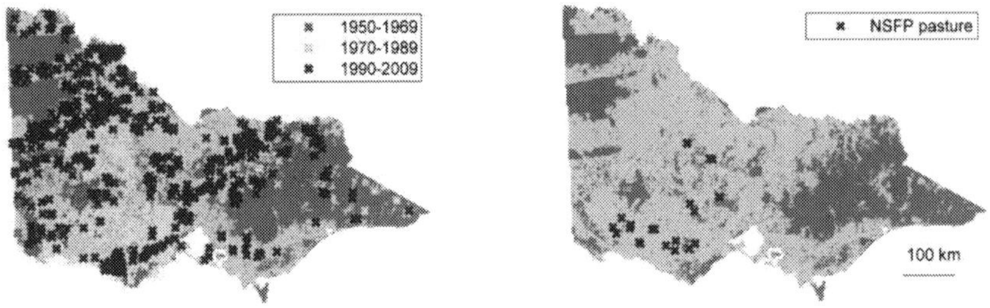

Figure 1. (left) Locations of VSIS measurements of topsoil pH across Victoria sorted according to sampling date; (right) locations of NSFP pasture sites where pH topsoil is measured in 1970-72, 1993 and 2011-13. Public land is coloured dark grey.

2.4. Farmer requested soil analyses from Victoria Australia

MacLaren et al. [5] collated a dataset of 75 000 soil analyses requested by farmers in Victoria between 1973 and 1993. For each analysis, the soil submitted to the laboratory was a bulked sample judged to be representative of the 0-10cm layer of a particular paddock. The bulked soil samples were air dried and sieved to less than 2mm and then soil pH was measured in water with a soil to water ratio of 1:5. The site information recorded included the year of sampling, the nature of the next enterprise scheduled to use the paddock (pasture, horticulture, cereal, row crops or forestry), and the nearest location. The locations were geo-referenced using the Government of Victoria's gazetted place names. Since the precise coordinates of the sampled area in the field, are unknown it is not possible to pair observations made at the same location at different times.

The sampling intensity and the enterprises requesting tests varied in both time and space (Figure 2) since they are controlled by the requirements of farmers and land managers rather than statistical considerations. However, in comparison to VSIS there is reasonable spatial and temporal coverage of samples from pastures across Victoria (Figure 2a & 2c). The only possible exception to this is in the northwest of the state where crops are more prevalent. Cereal soils were only widely sampled from the late 1980s and these samples were predominantly taken from the northwest of Victoria (Figure 2b).

Soil tests are likely to have been requested for one of three reasons: (i) to optimize fertilizer additions (ii) to identify reasons for poor plant growth (iii) to monitor soil quality. The reason for requesting a particular test is not recorded and there is potential for soils portraying particular characteristics (e.g. poor plant growth) to be preferentially sampled. No tests were requested from public land which is coloured grey in figure 2.

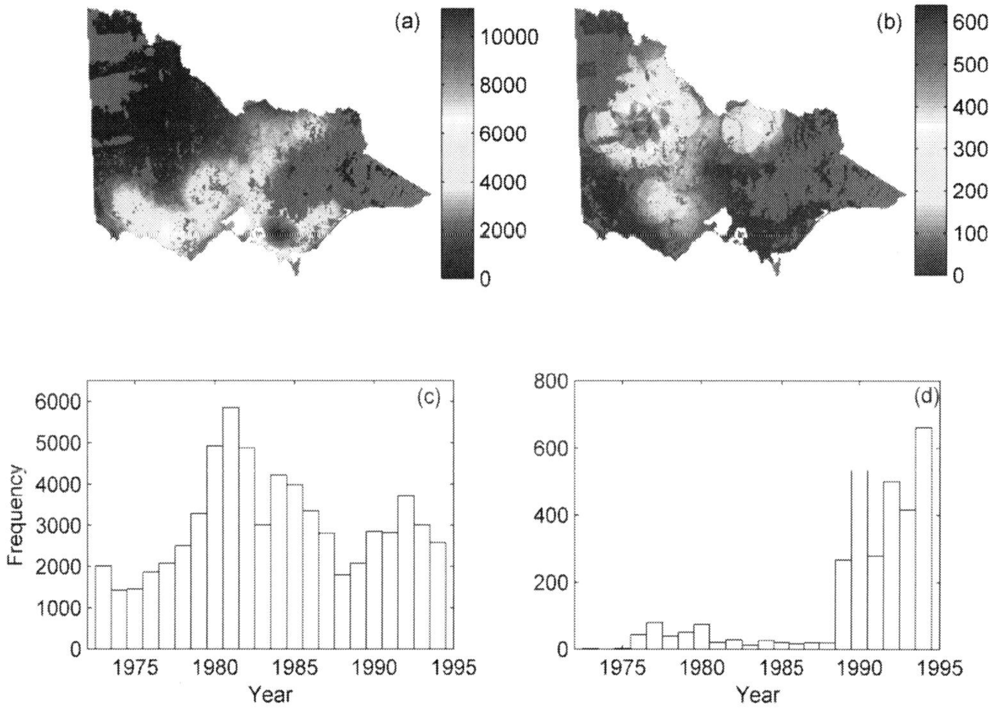

Figure 2. (a) Number of pasture soil samples requested within 50 km; (b) number of cereal soil samples requested within 50 km (c) histogram of pasture soil samples each year; (d) histogram of cereal soil samples each year.

2.5. *Primary Production Landscapes*

An important consideration in any spatial study of a soil property is the support size or coverage of each prediction of the property. For example the measurements within the farmers' dataset each had a support of a single paddock. It might be considered natural to therefore report predictions of topsoil pH with the same support. However, it is questionable whether this is the most meaningful support size given that different schedules of soil management practices such as liming are likely to lead to large variations in pH for paddocks in the same locality. These variations due to management schedules could hide the variations due to larger-scale factors such as climate. At the other extreme, one might predict the temporal variation of the mean topsoil pH value across the whole state. However Victoria covers an area of 227 000 km^2 and contains vastly different soil types and climatic conditions (Figure 3b) and therefore a single state-wide value would fail to recognize important regional variations. MacEwan et al. [7] addressed this problem by dividing Victoria into 22 Primary Production Landscapes (PPLs; Figure 3a). These landscapes were regions of relatively constant climate, dominant soil types and management practices and were considered suitable areas over which variability in agricultural productivity could be assessed and advice could be given.

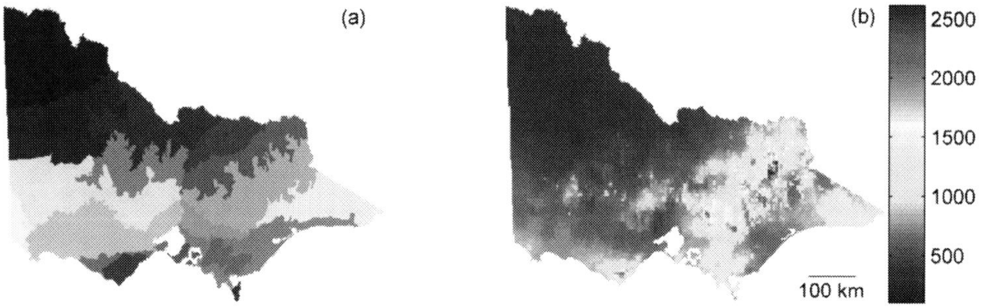

Figure 3. (a) The 22 PPLs of Victoria as designated by MacEwan et al. [7]; (b) mean annual rainfall (mm) across Victoria between 1973 and 1994.

2.6. Overview of this paper

We are interested in the state-wide mapping of topsoil pH status and its variation in time and space. We have seen that the data contained in VSIS tend to be spatially clustered for each time period. Thus these data are not suitable for state-wide mapping since variations in time are confounded with variations in space. There is much more potential to use these data to quantify the status at the same or a smaller spatial scale as the original projects which contributed to the database. In cases where sites or regions have been re-visited it is also possible to assess the local change in pH that has occurred between visits. Examples of such re-visits within the database are rare but new survey efforts could be arranged at the sites of VSIS soil sites. Crawford and Robinson [8] describe a pilot study which uses this approach to assess change in topsoil pH at sites in the west of Victoria where pH was measured between 1968-72 as part of the National Soil Fertility Project (NSFP; [9]). Crawford et al. [10] had previously re-measured topsoil pH in 1993 at 31 NSFP sites (Figure 1 right). A further possible use of the VSIS data is to validate models of soil variation inferred from other data sources. In particular the NSFP data were collected and analyzed according to a similar protocol as the farmers' data.

In this paper we focus on the farmers' data collected from pastures since these data have good spatial and temporal coverage between 1973 and 1994. Preliminary analyses suggest that pH measurements amongst the farmers' data that were collected from cereal enterprises were significantly higher than those collected from pastures in the same locality. Therefore we do not combine different enterprise types within our analyses. We describe the statistical models that are required to account for the uneven sampling intensity evident in figure 2. Then we present predicted maps of topsoil pH status in 1973 and 1994 and the change in pH that occurred between these dates across Victoria. Finally we discuss the extent to which these predictions can be trusted and the model validation that is required. This includes validation using the NSFP data.

3. Methods

3.1. Spatio-temporal modelling of the Victorian Farmers' data

A statistical model is required to infer trends in the observations gathered from any non-probabilistic or purposive sample design [6]. Figure 2 shows that this issue is particularly pertinent for the farmers' dataset since the samples are unevenly distributed across Victoria. Therefore a linear mixed model (LMM) of the space-time correlation between the pH observations within the farmers' dataset was estimated. Note that such a model can account for the uneven distribution of samples in space and time. It cannot account for any preferential sampling (e.g. if a farmer is more likely to request an analysis of fields he expects to be acidic).

The complete farmers' dataset consists of 75 000 observations. Such a large dataset can be cumbersome to analyze and therefore we focus on measurements made at the start and end of the survey period. We treat the $n_1 = 4\,125$ measurements made between 1973-75 (time period t_1) and the $n_2 = 9\,307$ measurements made between 1992-94 (time period t_2) as two spatially correlated variables denoted z_1 and z_2 respectively. The location recorded for each observation corresponds to the nearest gazetted place name. There were $n_p = 1\,988$ unique combinations of place name and time periods at which pH observations were recorded. The LMM is written:

$$\begin{bmatrix}\mathbf{z}_1\\ \mathbf{z}_2\end{bmatrix} = \mathbf{M}\boldsymbol{\beta} + \mathbf{DU} + \boldsymbol{\varepsilon}. \quad (1)$$

Here $\mathbf{z}_1$ and $\mathbf{z}_2$ are the vectors of observations of z_1 and z_2. $\mathbf{M}$ is a fixed effects design matrix. It consists of q columns and is of length $n = n_1 + n_2$. Each column of $\mathbf{M}$ contains a factor which explains some of the variability in topsoil pH. $\boldsymbol{\beta}$ is a length q column vector of fixed effects coefficients. In our model, $q = 2$. The first column of $\mathbf{M}$ consists of ones and the first entry of $\boldsymbol{\beta}$ corresponds to the space-time mean of topsoil pH across Victoria within the two time periods. The second column of $\mathbf{M}$ contains the natural logarithm of the mean annual rainfall between 1973 and 1994 at the site of each observation and thus the fixed effects are linearly related to this factor. Note that the fixed effects vary in space but not time. This is to ensure that the fixed effects did not lead to temporal trends within the predicted maps in regions where there were too few data to infer this trend. $\mathbf{D}$ is a $n \times n_p$ fixed effects design matrix. For $i = 1, \ldots, n$, the ith row of $\mathbf{D}$ contains a 1 in the column corresponding to the place and time period combination of observation i. The other entries of the row are zero. $\mathbf{U}$ is a length n_p vector which contains the spatially and temporally correlated random effects associated with each place and each time period combination. The entries of $\mathbf{U}$ are realizations of a Gaussian random function with mean zero and covariance matrix $\mathbf{C}$. Finally, $\boldsymbol{\varepsilon}$ is a vector of independent and identically distributed Gaussian random effects associated with each observation. The inclusion of this term permits different observations recorded at the same location and in the same time period to have different values. The entries of $\boldsymbol{\varepsilon}$ have mean zero and variance σ_ε^2.

The covariance matrix of the correlated random effects, $\mathbf{C}$, was determined from a linear model of coregionalization (LMCR) of the random effects of z_1 and z_2 (see [11]). This model has three components: an autovariogram which describes the spatial dependency amongst the random effects for measurements made in period t_1; an autovariogram which describes the spatial dependency amongst the random effects for measurements made in period t_2 and a cross-variogram which describes the spatial dependency between the random effects of an observation made at time $t_1$and the random effects of an observation made at time t_2. The parameters of the LMCR and the independent variance parameter σ_ε^2 were fitted to the available data and by the method of moments [12]. Matérn models, [13], were used to represent each variogram.

When we use the LMM to simulate soil pH measurements (see Section 3.3) we assume that the random effects of each variable are a realization of a Gaussian random function. The observed data were skewed (skewness=1.93) and inconsistent with this assumption. Therefore prior to model estimation the observed pH measurements were log-transformed and shifted and this reduced the skewness to 0.7 (Figure 4).

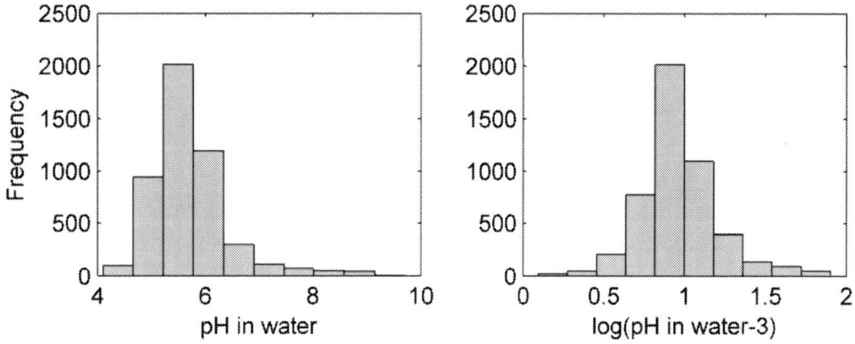

Figure 4. (left) Histogram of 1973 measurements pH in water on pasture; (right) histogram of 1973 log (pH in water – 3) on pasture.

3.2. Validation of the linear mixed model
The estimated LMM was validated by leave-one-out cross-validation. In this procedure an observation is removed from the dataset and the remaining data are used to predict that observation by co-kriging [12] and to calculate the co-kriging variance for that prediction. Then the standardized prediction error – the squared difference between the predicted value and the observed value all divided by the co-kriging variance – is calculated. The procedure is repeated removing each observation in turn. If the prediction errors are normally distributed then the standardized prediction errors will be realizations of a first order chi-squared distribution. This distribution has mean 1.0 and median 0.45. The farmers' data and the fitted LMM were also used to predict topsoil pH at the NSFP pasture sites shown in figure 1. These predictions were compared with the values observed in the NSFP survey.

3.3. Mapping the status and change in topsoil pH
The estimated LMM of (transformed) topsoil pH was used to simulate 1000 realizations of z_1 and z_2 at the nodes of a regular 2 km grid which covered Victoria using the conditional LU simulation algorithm [12]. Each simulated value was back-transformed and the average simulated pH value at each node in each time period and the average change in pH values between time periods at each node were calculated. The use of a stochastic simulation algorithm meant that the probability of the status or change exceeding a particular threshold could be extracted from the histograms of the simulated values. Each realization was up-scaled to the PPL-scale by calculating the average of the simulated values within each PPL.

4. Results
The auto and cross variograms of the LMM of transformed pH are shown in Figure 5. Substantial spatial correlation in transformed pH measurements is evident at distances up to 100 km. The estimated β value corresponding to the log rainfall factor amongst the fixed effects was -1.07 ± 0.16 log-pH units which indicated a significant and negative correlation between rainfall and soil pH. The independent random effects contributed 33% of the total variance of the transformed pH in the 1970s and 37% of the total variance in the 1990s.

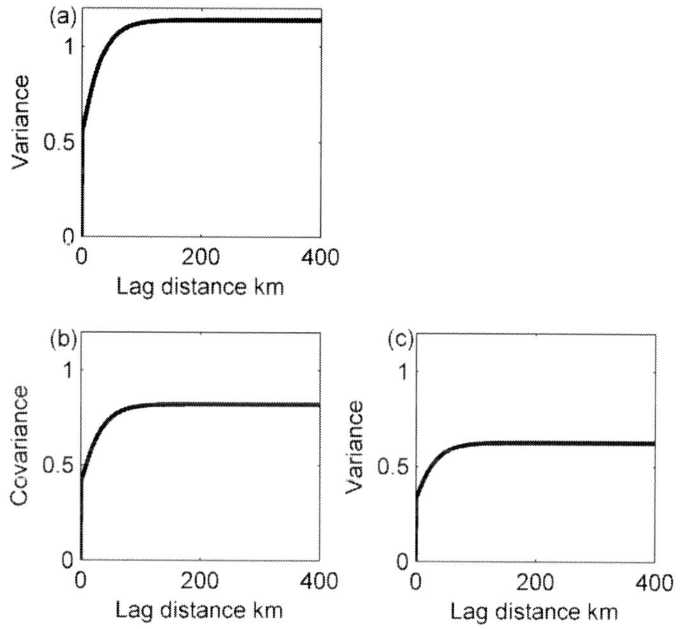

Figure 5. (a) Auto-variogram for transformed pH on pastures in 1973-75; (b) cross-variogram for transformed pH on pastures in 1973-75 and 1992-94 and (c) auto-variogram for transformed pH on pastures in 1992-94.

Clear spatial trends are evident in the maps of pH status on pastures in the early 1970s and early 1990s (Figure 6). The lowest pH values are in the highest rainfall areas. A substantial and increasing proportion of the state (23% in the early 1970s and 29 % in the early 1990s) has expected topsoil pH<4.8 implying that large areas of soil have acidified. The expected pH is greater than 4 at all locations in the Victoria in both time periods. However, the local variation between paddocks means that in each time period an average of 2% of the paddocks in each realization had pH < 4 and therefore have productivity compromised by extremely acidic soil (Figure 7).

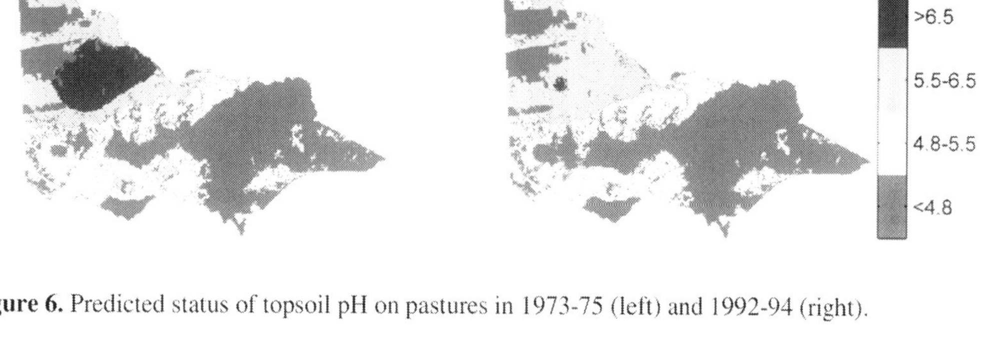

Figure 6. Predicted status of topsoil pH on pastures in 1973-75 (left) and 1992-94 (right).

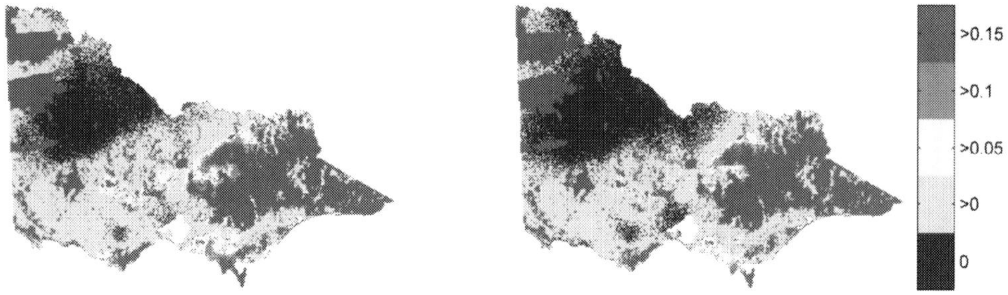

Figure 7. Map of the probability of a paddock having pH<4 in 1973-75 (left) and 1992-94 (right).

The expected change in topsoil pH on pastures over the duration of the farmers' data survey is negative for the majority of Victoria (Figure 8). Figure 8 (right) assesses where the predicted changes are statistically significant. The changes recorded in figure 8 (left) are the average change that is expected to have occurred within a paddock at a particular site. However this change will be uncertain. This uncertainty can be assessed by looking at the range of different changes that were recorded for the site amongst the 1000 simulated realizations. The dark blue portions of the map signify where more than 90% of the simulated realizations recorded an increase in soil pH during the course of the survey. Similarly the red portions signify where 90% of realizations recorded a decrease in soil pH. There is a small region in the southeast of the state where the probability of an increase is greater than 0.9. There are other small regions in the southwest where the probability of a decrease in pH is greater than 0.9. There is also a large region in the northwest where a decrease appears to be similarly certain. However it should be recalled that the samples were relatively sparse in this region.

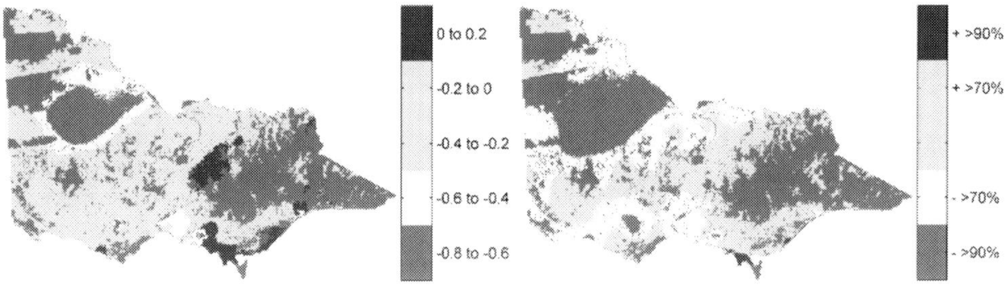

Figure 8. (left) Predicted change in topsoil pH on pastures from 1973-75 to 1992-94 at the paddock-scale; (right) Percentage of simulations where a local increase occurred '+' or decrease occurred '–' in topsoil pH on pastures at the paddock-scale.

The support of the predictions in figure 8 is a single paddock. Hence the local variation between paddocks means there are relatively few areas where a significant change is predicted. When this prediction support is increased to the PPL this local variability is smoothed over the larger area and the probability of a decrease in pH is greater than 0.9 for 11 of the 22 PPLs (Figure 9).

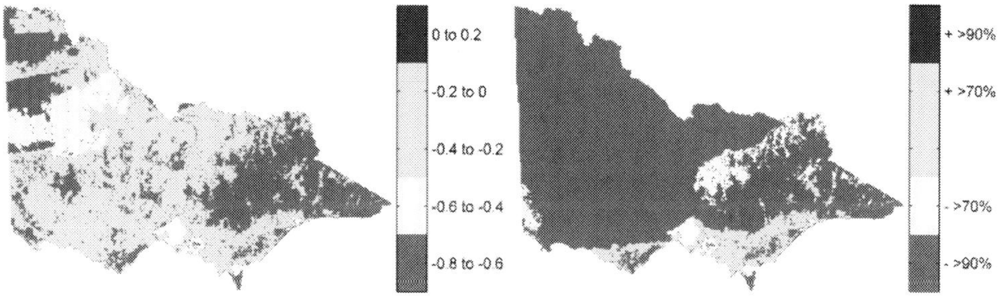

Figure 9. (left) Predicted change in topsoil pH on pasture between 1973-75 and 1992-94 at the PPL support; (right) likelihood of increase '+' or decrease '–' in topsoil pH on pasture at the PPL support.

When the same modelling procedure was applied to the farmers' samples from cereal enterprises rather than pastures a similar pattern of pH variation was evident (Figure 10). However pH values were larger in the croplands in the northwest of Victoria. The map of topsoil pH status for cereal enterprises in the 1970s (Figure 10 left) was based on a relatively small number of observations. Therefore this map is uncertain and no statistically significant temporal trends in pH for cereals were observed.

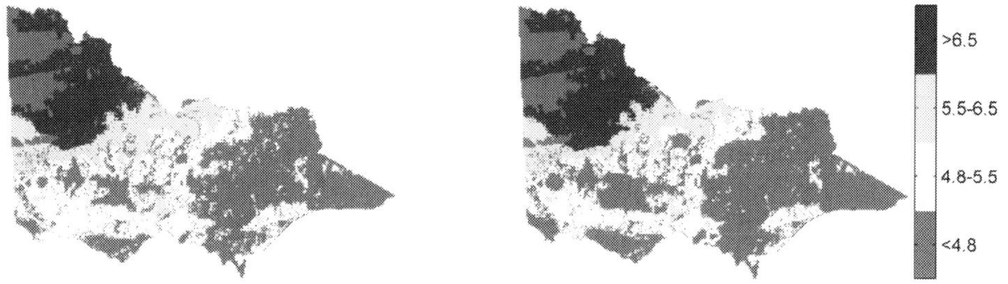

Figure 10. Predicted status of topsoil pH under cereals in 1973-75 (left) and 1992-94 (right).

When cross validation of the LMM for pastures was performed the mean and median standardized prediction errors were reasonably close to their expected values (mean =1.07; median =0.35). Figure 11 demonstrates that the LMM reliably predicted the spatial trends in the NSFP observations made in each of the two time periods.

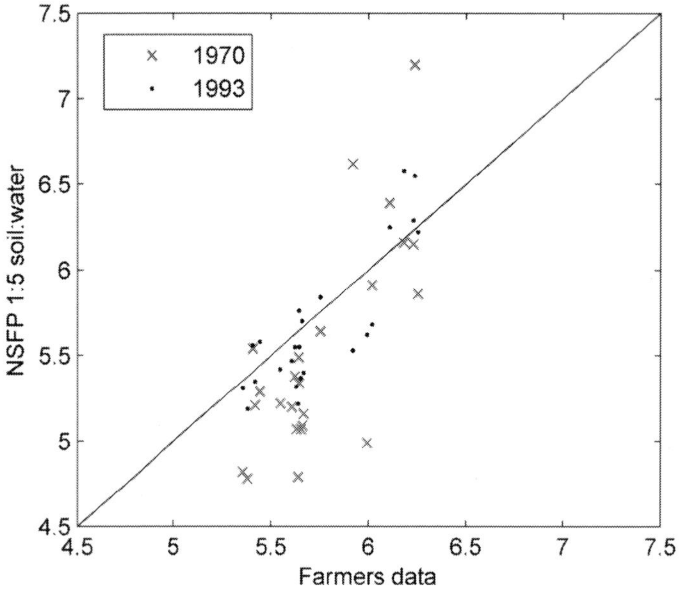

Figure 11. Validation of model of topsoil pH at NSFP pasture sites.

5. Discussion and Conclusions

In this paper we have seen evidence that the levels of acidity measured in farmers' soil samples in Victoria increased between 1973 and 1994. According to our space-time model of the variation of topsoil pH, in the early 1970s samples from 23% of Victoria were expected to have a pH of less than 4.8. This percentage increased to 29% by the early 1990s. Such levels of acidity are known to lead to severe productivity losses. The space-time models lead to maps which identify where the acidity threat is emerging and can be used by policy makers to focus remediation efforts.

Although the expected topsoil pH was never less than 4, the large component of local variation meant that in some areas the probability that paddock-scale acidity would exceed this threshold was as high as 0.2. Thus broad-scale soil monitoring can only be used to look at underlying trends in properties. Further soil sampling is required to inform paddock- and farm-scale soil management decisions.

Leave-one-out cross-validation suggests that the space-time models adequately represent the variation of measured pH values. However we have stressed the need to apply caution when interpreting these results. It is vital to consider whether the trends observed in the farmers' samples are necessarily representative of the soils of Victoria because some farmers may have preferentially sampled soils with particular characteristics. For example a farmer might focus his sampling in fields where he thinks that pH is changing rapidly or where he suspects it might be limiting crop production. Also, laboratory methods might evolve during the course of the survey and the absence of exact locations for each sample hinders the modelling. Such issues cannot be identified by internal cross-validation exercises. Validation exercises which use reliable data from other sources (e.g. the NSFP measurements) can give a better indication of whether the model predictions are consistent with the actual variation of Victorian soils. Such independent validation data is unlikely to be available throughout the state and often it is necessary to use expert opinion to validate the predicted maps. Experts might identify regions where the predicted behavior does not conform to their expectations. For example, the large decreases in pH on pastures in the northwest of Victoria were surprising.

Further examination of the predictions suggested that the large decreases might be artefacts caused by there only being a small number of gazetted place names for samples from this area and by some of these locations changing during the course of the survey. Such artefacts are unlikely to have substantial effects further south where the sampling is much more intensive and consistent in time and space.

The results of purpose-built SMNs might be expected to be more reliable than studies which use legacy data. SMNs with a probabilistic design do not require a model to quantify the variability amongst the data. Although a model is required for purposive designs we can generally be confident that these SMNs have not been preferentially sampled. That is not to say that the interpretation of purpose-built SMNs is not without challenges, for example the study of Chapman et al. [14] demonstrated the difficulties in ensuring that soil measurements made decades apart were equivalent.

Using legacy data to monitor soil does have its advantages. It is generally cheaper and quicker to yield results than a newly implemented purpose-built SMN. Often legacy data is the only source of information about the past state of soils. However this convenience comes at a cost since it is not generally possible to control the design of legacy datasets and therefore they might not be suitable to monitor a particular property of interest. For example we found that the VSIS data did not have sufficient spatial and temporal coverage to permit state-wide monitoring and the farmers' data only covered the period 1973-94 and had insufficient samples to monitor pH on croplands. Furthermore legacy datasets often fail to document exactly how the sampled locations were chosen which can hinder the correction of any sampling biases. If the exact locations are not recorded then it is not possible to identify cases where the same location has been re-sampled and to use these measurements from the same site as a direct measurement of change. Also the legacy data might require pre-processing before they can be used. For example the collation of the farmers' data took 15 months since handwritten reports had to be digitized.

Some countries are currently implementing more carefully organized soil monitoring efforts using farmer requested soil analysis results [3]. In France, analytical laboratories are given financial incentives to provide data. In such a monitoring effort there is potential to have more control over the collected data through requests to the laboratory to collect information on the reason why an analysis was requested, to remove measurements which result from preferential sampling for a particular soil attribute and to provide information on the exact laboratory protocols used. Also it might be possible to record when the same field has been resampled. Collecting data on management schedule could also be undertaken if this is provided with the documents accompanying the submitted sample. Critically this would provide an opportunity to elucidate the effects management on soil tests and more broadly inform on the effect of policy makers' and farmers' responses to the information gained from monitoring.

Acknowledgements

This research was enabled by funding from the Victorian Government through the Visiting Fellows program of the Department of Environment and Primary Industries and support for the 'Understanding Soil and Farming Systems' project. BPM gratefully acknowledges the OECD which funded his attendance at the 'Soil Change Matters' workshop in March 2014 through the Co-operative Research Programme: Biological Resources Management for Sustainable Agricultural Systems. BPM's contribution is published with the permission of the Director of the British Geological Survey (NERC).

References

[1] Kirk G J D, Bellamy P H and Lark R M 2010 Changes in soil pH across England and Wales in response to decreased acid deposition *Global Change Biology* **16** 3111-3119

[2] European Commission 2006 Proposal for a Directive of the European Parliament and of the Council. Establishing a Framework for the Protection of Soil and Amending Directive 2004/35/EC.COM(2006)

[3] Arrouays D, Saby N P A, Swiderski C, Chauveau D, Marchant B P, Orton T G, Lemercier B, and Walter C 2014 A soil test database for French agricultural soils. State of progress, advantages and limits for monitoring soil changes. Proc. of *Soil Change Matters* workshop, Bendigo, Australia, March 2014
[4] Environment and Natural Resources Committee 2004 Inquiry on the Impact and Trends in Soil Acidity. Parliamentary Paper No. 59 of session 2003-2004. Parliament of Victoria, Australia.
[5] MacLaren G S, Crawford D M, Brown A J and Maheswaran, J 1996 Temporal and spatial changes in soil chemistry: (I) pH and electrical conductivity In 'Australian and New Zealand National Soils Conference 1996: Soil Science - Raising the profile, 1st - 4th July 1996 University of Melbourne' (Melbourne, Australia: Australian Soil Science Society)
[6] Arrouays D, Marchant B P, Saby N P A, Meersmans J, Orton T G, Martin M P, Bellamy P H, Lark R M and Kibblewhite M 2012 Generic issues on broad-scale soil monitoring schemes: A review *Pedosphere* **22** 456-469
[7] MacEwan R, Robinson N, Imhof M, Rees D, Sposito V and Elsley M 2014 Primary Production Landscapes of Victoria *Proc. of 14th Australian Agronomy Conf.* http://regional.org.au/au/asa/2008/poster/agronomy-landscape/6250_macewanrj.htm
[8] Crawford D M and Robinson N 2014 A pilot survey of change in soil pH using legacy sites. Proc. of *Soil Change Matters* workshop, Bendigo, Australia, March 2014.
[9] Colwell J D 1977 'National Soil Fertility Project. Vol. 1, Objectives and procedures.' (Adelaide, South Australia: CSIRO Division of Soils)
[10] Crawford D M, Baker T G, and Maheswaran J 1994 Soil pH changes under Victorian pastures *Aust J Soil Res* **32** 105-115
[11] Marchant B P and Lark R M 2007 Estimation of linear models of coregionalization by residual maximum likelihood *Eur J Soil Sci* **58** 1506-1513
[12] Webster R and Oliver M 2007 *Geostatistics for Environmental Scientists.* 2nd Edition, (United Kingdom: John Wiley and Sons Ltd Chichester)
[13] Marchant B P and Lark R M 2007 The Matérn variogram model: Implications for uncertainty propagation and sampling in geostatistical surveys *Geoderma* **140** 337-345
[14] Chapman S J, Bell J S, Campbell, C D, Hudson G, Lilly A, Nolan A J, Robertson A H J, Potts J M and Towers W 2013 Comparison of soil carbon stocks in Scottish soils between 1978 and 2009 *Eur J Soil Sci* **64** 455–465

Assessing changes in carbon stocks of Scottish soils: lessons learnt

A Lilly and S J Chapman

James Hutton Institute, Craigiebuckler, Aberdeen, Scotland

E-mail: allan.lilly@hutton.ac.uk; steve.chapman@hutton.ac.uk

Abstract. Between 1978 and 1988, the soils at 721 locations throughout Scotland were sampled at intervals 10 km apart. They were described, characterised and samples were taken from each of the main horizons. Material not used for analysis was stored in the National Soils Archive. Between 2007 and 2009, 183 of these locations were re-visited (20 km intervals) and fresh samples taken to identify changes in nutrient status, pH and, in particular, soil carbon concentrations and stocks over the 19-31 year period. The archived soil samples from this time were re-analysed alongside those from the recent sampling to determine carbon concentrations. Near Infrared (NIR) spectroscopy was used to estimate their bulk density so that carbon stocks could be calculated as this was not measured at the original sampling. The results showed no statistically significant change in soil carbon stocks to 1 m depth for the main, broad land use types in Scotland apart from a small but significant increase (P=0.035) in soils under woodland. There was approximately 11.5% difference in carbon concentration between the re-analysed, archived soil and values originally obtained, but this was attributed to an artefact of differences in analytical methods. Between the two sampling periods, a decrease in carbon concentration of 2.4 g kg^{-1} was detected in cultivated soils. However, a significant increase in topsoil thickness of 2.9 cm (P=0.024) was sufficient to compensate for these changes in arable soils such that there was no detectable change in carbon stocks. The work shows the value of soil archives and of measuring horizon thickness.

1. Introduction

The Scottish Government has set ambitious targets to reduce greenhouse gas (GHG) emission to 58% of 1990 levels by 2020 and to 20% of 1990 levels by 2050 as detailed in the Climate Change {Scotland} Act 2009. It is estimated that Scotland's soils hold around 3000 Mt C and clearly losses of this carbon due to climate or land use change could impact on the Government's ability to achieve these targets, as well as adversely affecting soil functionality and food security.

In order to establish if Scottish soils were indeed losing soil carbon, the Scottish Government commissioned a research project to evaluate changes in carbon stocks (and other soil properties) across the whole country. This project not only assessed changes but investigated how sampling methods could affect the overall conclusions.

2. Material and Methods

Between 1978 and 1988 an inventory of soils in Scotland was undertaken by locating and describing the soil profile at each 5 km intersection between the easting and northing grid lines. The major soil horizons in each profile were also sampled at each 10 km grid intersection, giving a dataset of 721 sampled profiles. These samples were analysed to determine the concentration of the major exchangeable anions and cations, nitrogen, phosphorus, pH as well as carbon. The results of these analyses were stored in the Scottish Soil Database and material not used for analysis was stored in the National Soils Archive.

Between 2007 and 2009, 183 of these locations were revisited on a 20 km sample frame, that is, 25% of the original sites. At each location, the sampling position was confirmed by matching site characteristics such as slope, rockiness and aspect with the previous description, and a soil profile similar to the original was identified and each of the major soil horizons was sampled. Composite samples from 0-15 cm were also taken over a 20 × 20 m area using a soil auger and single cores from 0-15 cm and from 0-5 cm were taken from the profile pit at each of the sites.

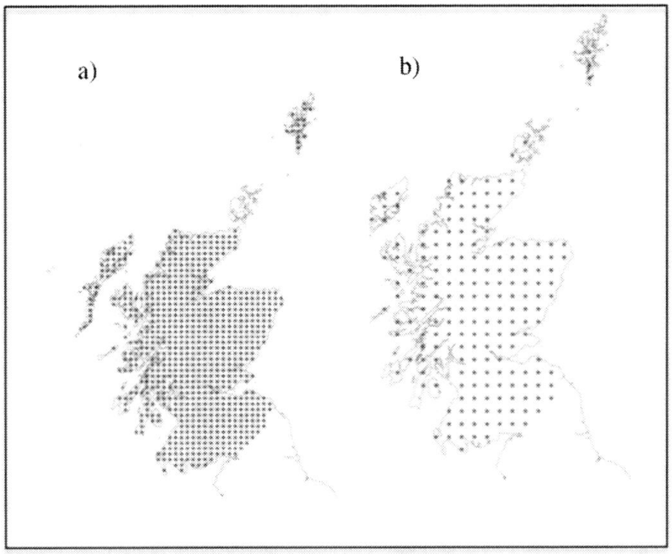

Figure 1: Distribution of 10 km (a) and 20 km (b) sample locations.

The subsequent analyses included those done previously as well as bulk density, a key property required in determining carbon stocks. In order to minimise instrument bias, the original archived soils were located and analysed alongside the newly sampled soils in randomised but paired batches. Samples for all analyses were initially air-dried at 30°C and then sieved (<2 mm). Samples for carbon content were further dried at 50°C and those for loss on ignition dried at 105°C. At each stage of subsampling, the soil was riffled to minimise sampling bias. The original, archive samples had also been air-dried at 30°C prior to storage.

Elemental carbon (%) for the original analysis was performed on a Hewlett-Packard CHN 185 analyzer but for the analysis of the new samples and for the reanalysis of archived samples, a Flash EA 1112 Series Elemental Analyser connected via a Conflo III to a DeltaPlus XPisotope ratio mass spectrometer (all Thermo Finnigan, Bremen, Germany) was used. The carbon concentrations were calculated from the area output of the mass spectrometer calibrated against a standard reference material which was analysed with every batch of ten samples [1].

The bulk density was determined for each major horizon in the soil profile for the 2007-9 soil samples, using, wherever possible, three replicates per horizon. Stainless steel cylinders, internal diameter 7.3 cm and height 5 cm, were pushed into the soil and then carefully removed, trimmed to remove excess soil and then extruded into a polythene sample bag, giving a soil volume rounded to 210 cm^3. These samples were then dried at 105°C, sieved (<2 mm) to remove stones and any roots and weighed. The bulk density of the fine fraction was obtained after making a correction for the volume of stones and roots.

Bulk density was not available for the 1978-88 soil samples so there was a need to estimate the bulk density for these soils in order to estimate carbon stocks. Many pedotransfer functions used to predict bulk density use carbon concentration as a predictor and using these pedotransfer functions would lead to a double reliance on the carbon concentration and subsequent error propagation.

To overcome these issues, Near Infrared (NIR) spectroscopy was used to predict soil bulk density. The measured bulk densities from the 2007-9 samples were used for calibration and subsequent prediction of bulk density values for both datasets to avoid the potential bias of having measured values for one set and predicted values for the other. NIR spectra of the <2 mm soil fraction (that is, the dried and sieved soil from the cores taken to determine bulk density and from the archived soil) were recorded from 1100 2500 nm at 2 nm intervals on a FOSS NIRS Systems 5000 spectrophotometer (FOSS NIRSystems, Silver Springs, MD, USA), using a transport module sampling attachment and a quarter cup sample holder. Reflectance mode spectra were collected using Infrasoft International ISIscan Software, Version 2.85.3 (FOSS Analytical AB, Hoganas, Sweden). Infrasoft International WINISI III Software, Version 1.50 E (FOSS Tecator AB, Hoganas, Sweden), was used for the development of calibration equations to predict the unknown bulk densities for the 1978-88 samples as well as predicting bulk densities for the 2007-9 soil samples

A total of 663 samples had measured bulk densities which were then correlated with the full spectrum of the NIR spectral data to develop an initial equation. A range of pre-processing methods, regression methods and different mathematical treatments were tried in order to obtain the best possible equation, which was validated by cross-validation. As the derived equation did not perform as well for highly organic samples (≥ 37% carbon) as for more mineral soil horizons, two separate regression equations were developed for soils with ≥37% carbon and <37% carbon, respectively [1].

The total profile carbon stock at each site, at each of the two sampling times, was calculated as the sum of the component soil horizon carbon stocks. The stock was calculated to 100 cm unless the profile was restricted by the presence of rock. Soils with flooding could not be sampled to depth. However, if the profile was flooded on one occasion but could be sampled on the other, the carbon stock was also calculated to 100 cm.

For each horizon, the carbon stock was calculated as the product of horizon depth, bulk density and carbon concentration with a correction made for stone content. As both carbon concentration and bulk density were measured on samples taken from the horizon midpoint, there was an assumption that they were representative of the whole horizon. In some cases, two (or three) samples were taken from the horizon, particularly for thick horizons; this was accounted for in the calculation by dividing the horizon into two (or three) equal parts and applying the corresponding sample values.

The comparison of the mean values for carbon concentration, bulk density, soil depth, stone content and carbon stock between soils sampled previously and the newer samples was done using a paired t-test with Genstat 13th edition (VSN International, Hemel Hempstead).

3. Results and discussion

The average time between samples was 25 years (19-31years) and we found no statistically significant change in soil carbon stocks to 1 m depth over this time period for the main, broad land uses types in Scotland (arable, improved grassland, semi-natural grassland, moorland and bog) apart from a small but significant increase (P=0.05) in soils under woodland (Table 1). However, this change was only apparent when the data for deep (>1 m) peat soils (Histosols) were removed from the analyses. The carbon concentration of Histosols varies little with depth, thus the main determinant in changes in carbon stocks in these soils is the overall thickness. In many cases the actual thickness of these soils could not be determined (the thickness exceeded the depth of the profile pit at around 1m and the additional probing with a 1 m-long soil auger) thus making the detection of change in these soils impossible within the scope of this resampling programme. Including these soils in the overall analyses only increased the mean carbon stocks making those sites where a real difference was detected less significant.

Table 1. Change in mean carbon stocks in Scottish soils over an approximate 25 year period by broad land use type.

Vegetation type	1978-1988 (C t ha^{-1})	2007-2009 (C t ha^{-1})	Change (C t ha^{-1})	No. sites[a]	*P* value
All	201.7 ± 11.7	209.8 ± 13.0	8.0	149	0.175
Arable	111.4 ± 16.5	111.5 ± 15.6	0.0	16	0.998
Improved grassland	123.5 ± 9.6	119.0 ± 9.9	-4.5	32	0.343
Semi-natural grassland	179.7 ± 25.1	185.2 ± 27.1	5.4	26	0.709
Woodland	163.2 ± 24.1	186.7 ± 26.9	23.5	21	0.035
Moorland	251.5 ± 20.7	257.7 ± 23.4	6.2	41	0.686
Bog	455.0 ± 34.8	489.6 ± 44.7	34.7	13	0.286

[a] deep peats (Histosols) and one soil derived from shelly sands removed from analysis.

However, interpreting the results was often far from straight forward and occasionally apparently contradictory. Although there was no change in carbon stocks of soils under arable cultivation, the carbon concentration in the Ap horizon of these soils decreased (36.0 to 33.7 g C kg^{-1}); however, there was a significant increase in topsoil thickness (Table 2) which meant that stocks remained the same (Table 1). Had the soils only been sampled for carbon concentration over 0-15 cm, as has been common in other UK sampling schemes [2, 3], we may have reported a carbon loss.

Table 2. Change in carbon concentration and horizon depth for Ap horizons in arable soils

Measure	1978-1988	2007-2009	Change	No. sites	*P* value
Carbon (g C kg^{-1})	36.0	33.7	-2.4 (-6.6%)	16	0.225
Horizon depth (cm)	28.9	31.9	2.9 (10.2%)	16	0.024

Carbon concentrations of the re-analysed, archived soil from the 1978-88 sampling were approximately 11.5 % less than the values obtained by the original analysis. The difference was not due to carbon loss during storage as a similar magnitude of loss was not found when we examined Loss on Ignition data. This suggests that the apparent difference in carbon concentration was due to differences in the protocols and the instrument used at that time (Figure 2a and b). Drying the samples before analyses may be a contributory factor and it is unclear how samples in 1978-88 were dried before analysis. Contemporary samples are dried at 50°C to avoid nitrogen losses. Weight losses after drying at 105°C compared to 50°C were 0.7% and 2.7% for a subset of mineral and organic soils, respectively, suggesting that a difference in drying regime could not explain the observed difference in carbon concentration.

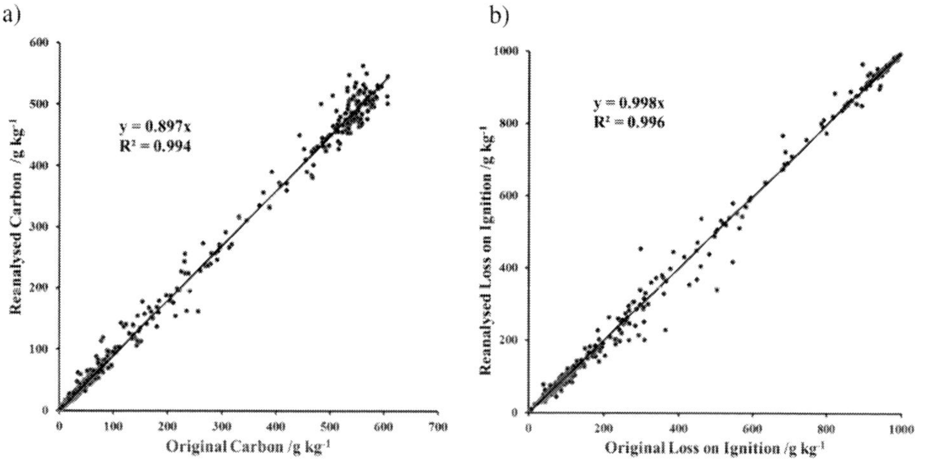

Figure 2: The relationship between the original measured carbon concentrations and the re-analysed soil (a) and the relationship between the original measured loss on ignition values and the re-analysed soil (b).

Having access to the archived material to re-run the analyses was a crucial element of the work, especially given the 11.5% change in carbon concentration that resulted from differences between the instruments used. Had this material not been available and we simply compared the results of the analyses over the two time periods then we would probably have drawn an erroneous conclusion that the carbon stocks of Scottish soils had declined by a considerable amount.

As mentioned, we could not determine carbon stocks fully for some deep peat soils. As the depth to the mineral layer or rock had not been recorded, there was an unknown stock of carbon below 1m, in some cases a considerable one. Hence these were removed from the statistical analyses.

As peat soils have a narrow range of carbon concentrations, the only way to determine changes in carbon stocks for these soils is to determine if the bog is growing or declining. While we were unable to do this for all peat bogs encountered at the sample locations, it is even more difficult to determine changes if the sample depth is 0-15 cm. Indeed, a formerly degraded bog could now be actively growing but as the bulk density of the upper layers would be less than that of amorphous peat, the stock calculation would suggest a decline.

Carbon stocks were calculated on the basis of equivalent volume but several authorities consider equivalent soil mass to be a better basis [4]. Using the equivalent volume may underestimate the true change in carbon, particularly where there is a significant change in bulk density. There is a particular problem in the case of organic soils; any increase in the depth of the organic layer will mean sampling less of the underlying mineral soil and vice versa for a loss of surface layer. However, it can be shown algebraically that even using soil mass (mineral plus organic) can under-estimate the change. Particularly for soils with significant organic content it is preferable to calculate on the basis of equivalent mineral weight. This assumes, of course, that there is no process causing loss or gain of mineral material. The idea is that the mineral material acts an anchor or reference point. We calculated the total mineral stock for our sites at both time periods. The data showed that for the more organic profiles there was an unacceptable degree of variability in the mineral content such that any recalculation based upon equivalent mineral mass would also give very variable carbon stocks. This was particularly applicable to soils under bog, moorland and woodland. For the more mineral soils under arable, improved grassland and most semi-natural grassland, this is an approach worth pursuing. Since our aim was to compare all the soils on an equal basis we have only presented results based upon equivalent volume, excluding the deep peats soils as mentioned above.

Another complicating factor in the analyses was the confounding factor of arable/grass rotations. While it was relatively easy to determine if the site was under continuous arable or permanent grassland, sites that had a mixed arable/grassland system could have been allocated to either arable or permanent grassland, often depending on the age of the grass. Results from an analyses of 2644 topsoils taken from the Scottish soils database, show significant differences ($p<0.01$) in mean carbon concentrations of soils under arable (3.66%), a mixed arable/grassland system (4.45%) and under permanent grassland (5.14%). Thus, the age of the grassland and its position in the rotation will influence the amount of carbon in the soil and its position in the rotation relative to the repeat sample could influence whether a decline or increase in carbon stock was detected. Better recording of land use history would help improve detection of change. Unfortunately, this is not always possible when sampling either randomly or on a grid. Selection of sites based upon known history may select for better overall management and thus bias the results.

Pairing sites gave increased statistical power to the analyses, since it reduced variability due to site differences. However, from a practical point of view, it was sometimes difficult to relocate and replicate the exact soil conditions. While external conditions such as slope, aspect and rockiness could be reasonably well matched, the variability inherent in soils, even over relatively short distances, meant that it could be difficult matching the sequence, thickness and presence of horizons exactly. Alternative approaches such as sampling on an offset grid, for example 1km east and north of the original, would have avoided these problems but would have meant many more sites would have needed to be visited in order to obtain a sufficient sample size. One method used to confirm that the resampled soils were similar to the original was to compare the Fourier Transform Infrared (FTIR) spectra of the mineral component of the soils (Figure 3). While the part of the spectra related to the organic component could change through time, for example, due to a land use change between the two sampling periods, the mineral component should remain the same. By overlaying the FTIR spectra, any differences in the mineralogy could be determined. In all but one case, a good match was found, confirming that these remaining sites were comparable.

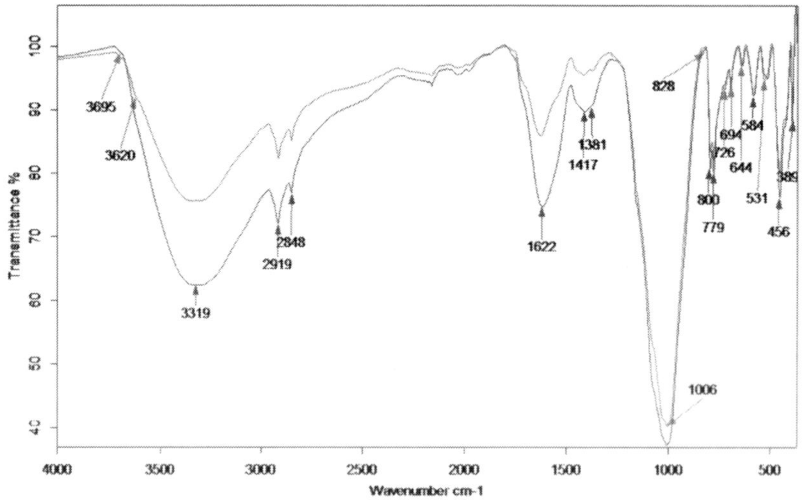

Figure 3: Example of the use of FTIR to confirm that soil sampled between the two time periods was comparable. The (lower) blue line is from the resampled topsoil (2007) and the (upper) red line is from the original (1987).

4. Conclusions

In conclusion, we were able to demonstrate that there has not been a significant loss in carbon from Scottish soils though interpretation of the results was not straight forward. The sample grid reflected the proportion of land uses in Scotland but also resulted in low sample numbers for some key land uses where carbon losses may be expected, for example, land under arable agriculture. Peat soils also posed a problem where the thickness exceeded the normal sampling depth of 1 m. In these soils, it would have been advantageous to determine a fixed datum such as the top of rock or drift underlying the peat so that changes in thickness could be recorded but even this has problems as bogs can shrink and swell depending on the degree of saturation.

Spectral analyses proved to be useful in both predicting bulk densities and in verifying that the site had been successfully relocated. This would not have been possible had the original samples from the 1978-1988 sampling campaign not been archived. The value of retaining archived material was also demonstrated by the difference in carbon values measured by two different elemental analysers. If the soil had not been available for reanalyses, there was a real danger of inferring that carbon was being lost from Scottish soils.

We have shown that sampling by horizon was superior to sampling by fixed depths particularly for cultivated soils but there is still an assumption that a sample taken from the middle of a horizon is representative of the horizon as a whole. While this may hold true for horizons that are regularly mixed, it may not be true for highly layered soils.

Acknowledgement

This work was funded by the Scottish Government and AL gratefully acknowledges the funding from OECD to attend the Conference. We would like to thank Jean Robertson for providing the spectral analyses.

References

[1] Chapman S J, Bell J S, Campbell C D, Hudson G, Lilly A, Nolan A J, Robertson A H J, Potts J M and Towers W 2013 Comparison of soil carbon stocks in Scottish soils between 1978 and 2009 *Eur. J. Soil Sci.* **64** 455-465

[2] Bellamy P H, Loveland P J, Bradley R I, Lark R M and Kirk G J D 2005 Carbon losses from all soils across England and Wales 1978-2003. *Nature* **437** 245-248

[3] Emmett B A, Reynolds B, Chamberlain P M, Rowe E, Spurgeon D, Brittain S A, Frogbrook Z, Hughes S, Lawlor A J, Poskitt J, Potter E, Robinson D A, Scott A, Wood C and Woods C *Countryside Survey: Soils Report from 2007.* Technical Report No. 9/07 (Wallingford, United Kingdom: NERC/Centre for Ecology and Hydrology)

[4] Ellert B H, Janzen H H and Entz T 2002. Assessment of a method to measure temporal change in soil carbon storage. *Soil Sci. Soc. Am. J.* **66** 1687-1695

Modelling soil carbon in agricultural systems: a way to widen the experimental space

Garry O'Leary[A], De Li Liu[B,C], James Nuttall[A], Muhuddin Rajin Anwar[B,C] and Fiona Robertson[D]

[A]Department of Economic Development, Jobs, Transport and Resources, PB 260 Horsham Vic, 3400, Australia. E-mail: garry.o'leary@ecodev.vic.gov.au; james.nuttall@ecodev.vic.gov.au

[B]Department of Primary Industries, Wagga Wagga Agricultural Institute, Wagga Wagga, NSW 2650, Australia E-mail: de.li.liu@dpi.nsw.gov.au; muhuddin.anwar@dpi.nsw.gov.au

[C]Graham Centre for Agricultural Innovation (an alliance between NSW Department of Primary Industries and Charles Sturt University), Wagga Wagga, NSW 2650, Australia,

[D]Department of Economic Development, Jobs, Transport and Resources, Mt Napier Rd Hamilton VIC 3300, Australia. E-mail: fiona.robertson@ecodev.vic.gov.au

Abstract. Mechanistic and explanatory simulation models provide robust and objective methods to extrapolate likely responses of crops and soils to climate change over different landscapes and time periods. Central to such simulation models are the supply of mineralised nutrients, in particular nitrogen, to crops through linked crop and nutrient sub-models that is achieved through modelling soil carbon dynamics. Attention to soil processes is therefore an essential part of building robust and sustainable production systems and understanding the potential impacts of climate change. To the farmer, focus must be on the productive capacity of the land and its rejuvenation to sustain production. In the broader context of reducing atmospheric CO_2 concentration through soil C sequestration, understanding soil processes and the immediate environment likewise require attention to productivity issues. This is because without maintaining productivity a better understanding of soil organic carbon (SOC) processes is unlikely to lead to increased SOC sequestration in Australia's farming land. Some gaps in knowledge of how to manage SOC are being addressed in a national research effort, including the scant measured data against which models can be tested. Nevertheless, continuing to apply models to push the boundaries well beyond what can be achieved in practice widens the experimental space, allowing new ideas to be tested where physical experiments are not possible. This raises optimism that new ways may be discovered to explain change in SOC and increase SOC where it is possible in a beneficial way.

1. Introduction
Modelling is widely applied to explore practicable and theoretical options in various scientific disciplines. The very early modelling attempts of engineers to model the flow of mass and/or energy [e.g. 1]) meant that nowadays the engineering disciplines like civil engineering, aerospace and mathematics have modelling as an integral part of any serious research enquiry [e.g. 2]. There are

other disciplines that are still catching up and there are others such as soil science, where there are many advanced models describing numerous components, that will allow their application well beyond what is presently seen.

Simulation models can provide robust and objective methods to extrapolate likely responses of crops and soils to climate change over different landscapes and time periods [e.g. 3]). Whilst simulation modelling can be easily confused with statistical descriptive modelling its distinguishing feature is its propensity for extrapolation beyond the original data. Central to such crop simulation models is the supply of mineralised nutrients to crops, in particular nitrogen, through linked crop and nutrient sub-models that is achieved through modelling soil carbon dynamics [e.g. 4, 5, 6, 7]). Carbon in the soil is primarily linked to carbon in the atmosphere through vegetation and animals. Free Air CO_2 Enrichment (FACE) experiments have shown that crop yields are increased at elevated atmospheric CO_2 concentrations [8], whilst crop nutrient levels (e.g. grain nitrogen concentration) are usually decreased [9, 10]. Such increase in crop growth is likely to result in an increase in soil organic carbon (SOC) storage without water stress but less likely under water stress [11]. Because of the critical nature of mineral nitrogen supplies for crops to maintain high productivity and the effects of immobilisation and mineralisation under altered SOC conditions a focus is required not only on the soil but also the vegetation management. Such a focus has been made in forest systems that have shown no evidence of a progressive nitrogen limitation (PNL) under a prolonged elevated CO_2 environment [12] although for broad acre production systems PNL may impact future N-supply to plants [13]. Therefore, attention to processes, rather than statistical associations, is an essential part of building robust production systems. To the land manager and farmer, focus must be on the productive capacity of the land and its rejuvenation to sustain production and ecosystems beyond the present. This is particularly the case for degraded soil if productivity is to be sustained.

In a broader context of reducing atmospheric CO_2 through soil C sequestration across large areas attention to productivity is considered crucial not only because if its direct link to increased biomass production but also because of the economics of production. The value of production and SOC and their biophysical relationship will therefore be important in our understanding and management of soil C sequestration in Australia's farming land. Modelling together the biophysical and economic constraints may also provide the opportunity to identify the best option to increase SOC within specific regions and agro-ecological zones. Some gaps in our knowledge of how to manage SOC are being addressed in a national research effort. A necessary part of this is the testing of models against the scant measured data that are available. But modelling offers more than testing our understanding of processes against measured data, it allows the exploration of management practices that may offer potential to increase SOC sequestration.

2. Measurements

The measurement of components of the soil involved in SOC dynamics is fundamental to any carbon modelling effort. Standards need to be set so that others can repeat the observations. Hence the national focus on standardised measurement is very helpful to advancing our knowledge and management of soil carbon.

A realistic representation of soils/treatments from soil sampling, often requires a composite sample, which is bulked from multiple sub-samples. It is clear that a large number of sub-samples provide a more realistic representation of the soils/treatment in the composite sample. Chan et al. [14] recommend at least 20 sub-sample cores for obtaining a representative composite sample for estimating the soil C at a particular site. In reality, due to the high costs of such field measurement, the number of samples to make a composite sample typically varies and often is much less than a recommended number, thus introducing more error particularly those relying on repeated measures. For example, in the same long term experiment, the number of sub-samples varied from four [15] to at least 10 [16, 17]. One of the problems in measured SOC data reported is its large variation that is attributed to many reasons including the lack of resource for conducting far more than the actual sub-samples taken. An example of this problem is taken from a well managed long term experiment

(Figure 1). Differences in SOC between two consecutive years can be up to several tonnes, the likely effect of inherent spatial variability, which can be over 30 times the mean rate of change (e.g. 250 kg/year) [18]. Such variation shows impossible mass balance between samplings times. Whilst increased sampling size and extending the sampling intervals by many years are recommended to preserve mass balance, modelling allows the exploration of hypotheses without such limitations.

Another problem in our experimental measurements of SOC to address soil carbon sequestration is the measurement and definition of the plant material that is returned to the soil. This is a key carbon input that drives the microbial processes for soil carbon turnover. Farmers need to understand the differences and ultimately the effectiveness of their SOC change from various sources of carbon input. Large differences exist between crop residues left over on the surface to those that are effectively incorporated into the soil by tillage or animals. Cropping residues are often estimated from a harvest index for running a simulation model, such as RothC. But this is only part of the carbon dynamics as plant material from above and below ground sources contribute. Although we can measure SOC with some degree of accuracy and defined uncertainty, the amount of plant root material is scarcely measured, instead estimating it from above ground production using an assumed root-shoot ratio [19, 20]. In addition we do not have the luxury of measuring what we would like because it is too costly. A well tested simulation model is clearly an efficient means of addressing many aspects beyond the possibility of experimental measurement and provides proof of concept for long-term management systems without being conflicted by cost and experimental error associated with repeated measures.

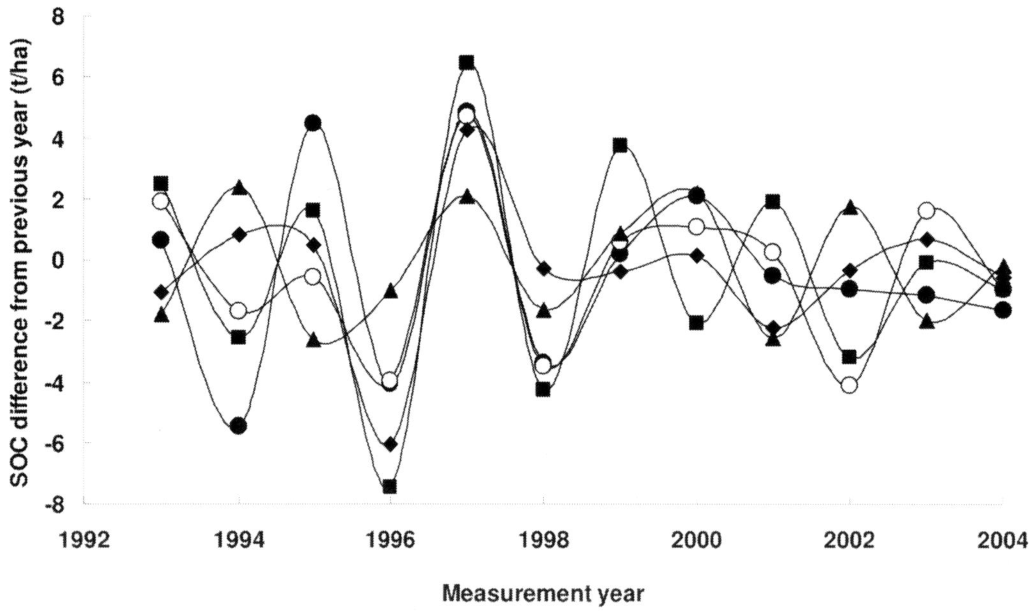

Figure 1: The difference in soil organic carbon (SOC) measured between two consecutive years from a long term cropping experiment (data reanalysed from Liu et al. [19]). No-tillage - stubble retained WL (●), Conventional cultivation - Stubble retained WL (○), No-tillage - Stubble burnt WL (▲), Conventional cultivation - Stubble burnt WL (■), Conventional cultivation - Stubble burnt WW (♦), WL: Wheat-lupin rotation, WW: Wheat-wheat rotation.

3. Modelling

Most soil carbon models are based on first-order decay of soil organic carbon with the role of microbes as decomposers controlling the decay rates [21]. Consequently, SOC is the result of the equilibrium balance of carbon inputs from plant-derived materials and outputs through decomposition affected by soil water and clay content and temperature. Many reports have demonstrated how various SOC models can simulate these processes against measured SOC [22, 23] or comparative modelling studies against measured SOC [e.g. 24, 25]). However, there are very few modelling studies that attempt to explore strategies that increase SOC or explain changes over time compared to those that rely on long term field experiments [e.g. 26]. Wang and Dalal [27] reported that only no till stubble retention with N application over 33 years long term experiment could increase SOC in Queensland, Australia. Conant et al. [28] reviewed the effect of pasture management and showed that SOC can be increased by improved management practices such as fertilisation, improved grazing management, sowing of legumes and grasses, earthworm introduction and irrigation. Chan et al. [29] used paired-site approach to compare several farming practice systems in south-eastern Australia and found significantly higher SOC stocks occurred only as a result of pasture improvement using P application over unimproved pastures. The changes in SOC were clearly influenced not only by farming practices, but also by climate and soil, and their interactions [30, 31]. It is therefore a challenge for scientists to take disparate experimental results and help farmers develop reliable practices to boost SOC in their paddocks.

The modelling of such inputs and outputs of C varies among models, but some account of important components such as soil erosion is important. Modelling offers the ability to explore practices that have not been robustly tested. For example, SOC levels in a long term experiment have been observed to decrease where stubble was retained on the soil surface in a semi-arid environment [32]. However, subsequent simulation analyses showed that surface retention of the stubble is only equivalent to 26% incorporation of stubble into the soil [19]. With this level of stubble return, SOC would vary widely over a spatial scale because of climatic and soil differences (Figure 2). In this way, modelling opened up a new line of enquiry with respect to developing ways to increase SOC. More recent work from Liu et al. [33] shows that it appears possible to increase SOC in rainfed wheat cropping soils of south eastern Australia by incorporating some higher threshold amount of wheat stubble into the soil. Whilst the suggested incorporation thresholds appear quite large at 50-70% of harvested stubble, practical methods need to be developed and tested to see if this can be achieved in the field. Liu et al. [33] goes on to argue that if such levels of incorporation can be achieved in practice then large gains could be made that potentially could assist Australia meet its emission reduction targets.

That modelling study widened the experimental space and proposed an optimistic approach to soil carbon management. Most long term agronomic experiments were never specifically designed to increase SOC and so don't tend to have very high C input levels, which might help us better understand the mechanisms. Therefore, the reliance on historical experiments not specifically designed to increase SOC may be problematic because statistical associations, whilst evident, need further articulation of the mechanisms involved in boosting SOC. Modelling employing mechanistic and explanatory variables offers a way to break free of this limitation, particularly with their ability to easily examine practices that have not been experimentally tested.

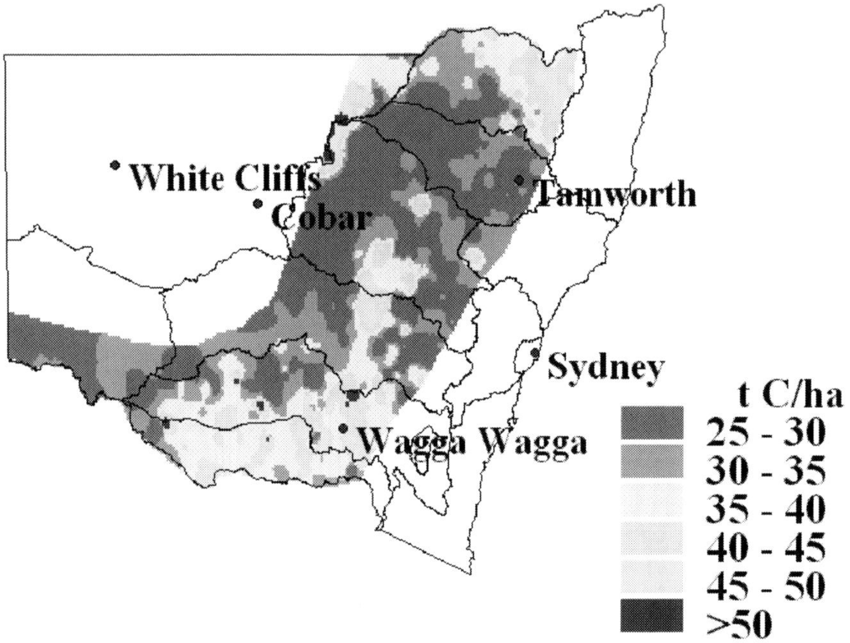

Figure 2: Simulated soil organic carbon (SOC 0-30 cm) equilibration level over New South Wales showing the large spatial variation in SOC from 25 to over 50 t/ha after 50 years under a present climate cropping regime with 26% residue incorporation (from Liu et al. [33]).

4. A way forward

Although modelling is a useful tool to widen the experimental space in soil science, we see that the performance of our contemporary models needs to be better documented. Typical measures of performance should include statistical measures of bias and relative and absolute accuracy such as root mean square error. Whilst the deficiencies in model design and sources of input data error will be apparent in many cases, the measured overall performance is an objective way to assess the extent that any particular model can be relied on and how much SOC change might be realistically achieved. If models are tested against measured data, then the accuracy of both model and data can be articulated and therefore advance soil carbon modelling. We suggest that long-term data sets involving SOC could be made more readily available in formats that could be rapidly applied to modelling studies.

There is a clear need in Australia to commence new long-term field experiments that attempt to increase SOC. Field experimentation examples that are pushing the boundaries are the work of Gill et al. [34, 35, 36] where organic material is placed into the soil with resultant large changes in soil properties and improved productivity. Farming systems that have high productivity in terms of the accumulation of biomass that can find its way into the soil are clearly ones that will feature more in future farming systems that raise SOC. Farming systems in the higher rainfall and cooler areas offer potential in this regard, and modelling such systems widen the experimental space for raising SOC levels in other parts of the Australian landscape. For degraded soils, there may be scope for increasing SOC, because they are starting from low levels and modelling can provide a way to explore options.

A wider experimental space will allow new untested ideas to be pursued where physical experiments are not possible. This raises optimism that new ways may be discovered to explain change in SOC and increase SOC where it is possible in a beneficial way.

Acknowledgements
This work was funded by the Victorian State Government Department of Economic Development, Jobs, Transport and Resources and the Australian Government Department of Agriculture as part of its Carbon Farming Futures Filling the Research Gap Program. Project 01203.013: Increasing soil carbon

in Eastern Australian farming systems: linking management, nitrogen and productivity. We thank an anonymous referee for helpful comments on an earlier manuscript.

References

[1] Forrester J W 1961 Industrial dynamics. (Waltham, MA: Pegasus Communications)
[2] NASA Modeling Guru 2014 https://modelingguru.nasa.gov/index.jspa. Internet address verified 5 May 2015 (NASA: USA)
[3] Christy B, O'Leary G, Riffkin P, Acuna T, Potter T, Clough A 2013 Long-season canola (Brassica napus L.) cultivars offer potential to substantially increase grain yield production in south-eastern Australia compared with current spring cultivars. *Crop Pasture Sci.* **64**, 901–913
[4] O'Leary G J, Connor D J 1996 A simulation model of the wheat crop in response to water and nitrogen supply: 1. Model construction. *Agr. Syst.* **52**, 1-29
[5] Reyenga P, Howden S, Meinke H, Mckeon G 1999 Modelling global change impacts on wheat cropping in south-east Queensland, Australia. *Environ. Modell. Softw.* **14**, 297-306
[6] Keating B A, Carberry P S, Hammer G L, Probert M E, Robertson M J, Holzworth D, Huth N I, Hargreaves J N G, Meinke H, Hochman Z, McLean G, Verburg K, Snow V, Dimes J P, Silburn M, Wang E, Brown S, Bristow K L, Asseng S, Chapman S, McCown R L, Freebairn D M, Smith C J 2003 An overview of APSIM, a model designed for farming systems simulation. *Eur. J. Agron.* **18**, 267–288
[7] Asseng S, Jamieson P, Kimball B, Pinter P, Sayre K, Bowden J, Howden S 2004 Simulated wheat growth affected by rising temperature, increased water deficit and elevated atmospheric CO_2. *Field Crops Res.* **85**, 85-102
[8] Leakey A, Ainsworth E, Bernacchi C, Rogers A, Long S, Ort D 2009 Elevated CO_2 effects on plant carbon, nitrogen, and water relations: six important lessons from FACE. *J. Exp. Bot.* **60**, 2859-2876
[9] Myers SS, Zanobetti A, Kloog I, Huybers P, Leakey A, Bloom A, Carlisle E, Dietterich L, Fitzgerald G, Hasegawa T, Holbrook N, Nelson R, Ottman M, Raboy V, Sakai H, Sartor K, Schwartz J, Seneweera S, Tausz M, Usui Y 2014 Increasing CO_2 threatens human nutrition. *Nature* **510**, 139–142
[10] Fangmeier A, Chrost B, Hogy P, Krupinska K 2000 CO_2 enrichment enhances flag leaf senescence in barley due to greater grain nitrogen sink capacity. *Environ. Exp. Bot.* **44**, 151-164
[11] Wood C W, Torbert H A, Rogers H H, Runion G B, Prior S A 1994 Free-air CO_2 enrichment on soil carbon and nitrogen. *Agric. Forest Meteorol.* **70**, 103-116
[12] Finzi A C, Moore D J, DeLucia E H, Lichter J, Hofmockel K S, Jackson R B, Kim H S, Matamala R, McCarthy H R, Oren R, Pippen J S, Schlesinger W H 2006 Progressive nitrogen limitation of ecosystem processes under elevated CO_2 in a warm-temperate forest. *Ecology* **87**, 15-25
[13] De Graff M, van Kessel C, Six J 2008 The impact of long term elevated CO_2 on C and N retention in stable SOM pools. *Plant Soil* **303**, 311-321
[14] Chan K Y, Oates A, Liu D L, Prangnell R, Poile G, Conyers M K 2010 A farmer's guide to increase soil organic carbon under pasture. 49pp. (Industry & Investment NSW, Wagga Wagga NSW Australia)
[15] Chan K Y, Heenan D P 1999 Lime-induced loss of soil organic carbon and effect on aggregate stability *Soil Sci. Soc. Am. J.* **63**, 1841-1844
[16] Chan K Y, Roberts W P, Heenan D P 1992 Organic carbon and associated soil properties of a Red Earth after 10 years of rotation under different stubble and tillage practices. *Aust. J. Soil Res.* **30**, 71-83

[17] Heenan D P, McGuhie W J, Thomson F M, Chan K Y 1995 Decline in soil organic carbon and total nitrogen in relation to tillage, stubble management, and rotation. *Aust. J. Exp. Agr.* **35**, 877-884
[18] Chan K Y, Conyers M K, Li G D, Helyar K R, Poile G J, Oates A, Barchia I M 2011 Soil carbon dynamics under different cropping and pasture management in temperate Australia: Results of three long-term experiments. *Aust. J. Soil Res.* **49**, 320-328
[19] Liu D L, Chan K Y, Conyers M K 2009 Simulation of soil organic carbon under different tillage and stubble management practices using the Rothamsted carbon model. *Soil Till. Res.* **104**, 65–73
[20] Liu D L, Chan K Y, Conyers M K, Li G, Poile G J 2011 Simulation of soil organic carbon dynamics under different pasture managements using the RothC carbon model. *Geoderma* **165**, 69-77
[21] Allison S D, Wallenstein, M D, Bradford, M A 2010 Soil-carbon response to warming dependent on microbial physiology. *Nat. Geosci.* **3**, 336-340
[22] Paul K I, Polgase P J 2004 Calibration of the RothC model to turnover of soil carbon under eucalypts and pines. *Aust. J. Soil Res.* **42**, 883–895
[23] Skjemstad J O, Spouncer L R, Cowie B, Swift R S 2004 Calibration of the Rothamsted organic carbon turnover model (RothC ver. 26.3), using measurable soil organic carbon pools. *Aust. J. Soil Res.* **42**, 79–88
[24] Parton W J 1996 The CENTURY model. In 'Evaluation of Soil Organic Matter Models.' (Eds D S Powlson, P Smith and J U Smith) pp. 283-291. (Springer-Verlag: Berlin)
[25] Smith P, Smith J U, Powlson D S, McGill W B, Arah J R M, Chertov O G, Coleman K, Frnko U, Frolking S, Jenkinson D S, Jensen L S, Kelly R H, Klein-Gunnewiek H, Komarov A S, Li C, Molina J A E, Mueller T, Parton W J, Thornley J H M, Whitmore A P 1997 A comparison of the performance of nine soil organic matter models using datasets from seven long-term experiments. *Geoderma* **81**, 153–225
[26] Lugato E, Berti A 2008 Potential carbon sequestration in a cultivated soil under different climate change scenarios: A modelling approach for evaluating promising management practices in north-east Italy. *Agr. Ecosyst. Environ.* **128**, 97-103
[27] Wang W J, Dalal R C 2006 Carbon inventory for a cereal cropping system under contrasting tillage, nitrogen fertilisation and stubble management practices. *Soil Till. Res.* **91**, 68-74
[28] Conant R T, Paustian, K, Elliott E T 2001 Grassland management and conversion into grassland: effects on soil carbon. *Ecol. Appl.* **11**, 343–355
[29] Chan K Y, Oates A, Li G D, Conyers M K, Prangnell R, Poile G, Liu D L, Barchia I M 2010 Soil carbon stocks under different pastures and pasture management in the higher rainfall areas of south-eastern Australia. *Aust. J. Soil Res.* **48**, 7-15
[30] Robertson F, Nash D 2013 Limited potential for soil carbon accumulation using current cropping practices in Victoria, Australia. *Agr. Ecosyst. Environ.* **165**, 130–140
[31] Badgery W B, Simmons A T, Murphy B W, Rawson A, Andersson K O, Lonergan V E 2014 The influence of land use and management on soil carbon levels for crop-pasture systems in Central New South Wales, Australia. *Agr. Ecosyst. Environ.* **196**, 147–157
[32] Heenan D P, Chan K Y, Knight P G 2004 Long-term impact of rotation, tillage and stubble management on the loss of organic carbon and nitrogen from a Chromic Luvisol. *Soil Till. Res.* **76**, 59–68
[33] Liu D L, Anwar M R, O'Leary G J, Conyers M K (2014) Managing wheat stubble as an effective approach to sequester soil carbon in a semi-arid environment: Spatial modelling. *Geoderma* **214-215**, 50-61
[34] Gill J S, Sale P W G, Tang C 2008 Amelioration of dense sodic subsoil using organic amendments increases wheat yield more than using gypsum in a high rainfall zone of southern Australia. *Field Crops Res.* **107**, 265–275.

[35] Gill J S, Sale P W G, Peries R R, Tang C 2009 Changes in soil physical properties and crop root growth in dense sodic subsoil following incorporation of organic amendments. *Field Crops Res.* **114**, 137–146

[36] Gill J S, Clark G J, Sale P W G, Peries R R, Tang C 2012 Deep placement of organic amendments in dense sodic subsoil increases summer fallow efficiency and the use of deep soil water by crops. *Plant Soil* **359**, 57–69

IOP Conf. Series: Earth and Environmental Science **25** (2015) 012018 doi:10.1088/1755-1315/25/1/012018

Soil carbon under perennial pastures; benchmarking the influence of pasture age and management

Susan E. Orgill[1], Nancy Spoljaric[2] and Georgina Kelly[3]
[1] NSW Department of Primary Industries, Wagga Wagga Agricultural Institute, NSW 2650, Australia.
[2] Monaro Farming Systems, PO Box 27, Bombala NSW 2632, Australia
[3] NSW Department of Primary Industries, Level 2, 470 Church St, North Parramatta, 2151, Australia.

E-mail: Georgina.Kelly@dpi.nsw.gov.au, Susan.Orgill@dpi.nsw.gov.au

Abstract. This paper reports baseline soil carbon stocks from a field survey of 19 sites; 8 pairs/triplet in the Monaro region of New South Wales. Site comparisons were selected by the Monaro Farming Systems group to demonstrate the influence of land management on soil carbon, and included: nutrient management, liming, pasture age and cropping history. Soil carbon stocks varied with parent material and with land management. The fertilised (phosphorus) native perennial pasture had a greater stock of soil carbon compared with the unfertilised site; 46.8 vs 40.4 Mg.C.ha to 0.50 m. However, the introduced perennial pasture which had been limed had a lower stock of soil carbon compared with the unlimed site; 62.8 vs 66.7 Mg.C.ha to 0.50 m. There was a greater stock of soil carbon under two of the three younger (<10 yr old) perennial pastures compared with older (>35 yr old) pastures. Cropped sites did not have lower soil carbon stocks at all sites; however, this survey was conducted after three years of above average annual rainfall and most sites had been cropped for less than three years. At all sites more than 20% of the total carbon stock to 0.50 m was in the 0.30 to 0.50 m soil layer highlighting the importance of considering this soil layer when investigating the implications of land management on soil carbon. Our baseline data indicates that nutrient management may increase soil carbon under perennial pastures and highlights the importance of perennial pastures for soil carbon sequestration regardless of age.

1. Introduction

Well-managed, perennial pastures may increase soil carbon (C) agricultural soils compared with cropping systems [1]. This is due to the extensive fibrous root systems of perennial pastures that may contribute more organic matter (OM) through root biomass and create drier soil conditions, and the minimal soil disturbance compared with most agricultural crops and cropping practices. In this study we used a paired-site approach to benchmark and compare; a) the age and management of perennial pastures on soil C stocks and b) the impact of changing land use from perennial pasture to annual cropping on soil C stocks. This paper presents baseline soil C (Mg.C.ha to 0.30 and 0.50 m) data for the Monaro region, New South Wales (NSW).

2. Methods

Site location and sampling

Nineteen sites were sampled in the Monaro region, southern NSW. The Monaro region is located 800 m above sea level with an average annual rainfall of 645 mm and is classified as Cfa - Köppen-Geiger climate classification [2]. Paired or triplet study sites were selected where the desired comparison was within 500 m. To be included in the paired comparison, sites were required to be on the same parent material (granite, sedimentary or basalt) and have similar soil and landscape attributes. Site comparisons were selected by the Monaro Farming Systems land-holder group to demonstrate the influence of management practices on soil C (Table 1). Sites with native perennial pastures (NPP) had never been cultivated and were typically wallaby grasses (*Rytidosperma spp.*), speargrasses (*Austrostipa spp.*) and snowgrass (*Poa sieberiana*). Introduced perennial pastures (IPP) were typically phalaris (*Phalaris aquatica L.*) and cocksfoot (*Dactylis glomerata L.*). Both NPP and IPP included exotic annual species such as subterranean clover (*Trifolium subterraneum*). Sites were sampled on a 25 x 25 m sampling grid according to SCRP protocols [3] using a hydraulic soil corer in late spring 2012. Ten cores were collected at each site to 0.50 m using a 40 mm diameter core and combined to form one composite sample for each soil layer (0.10 m increments) for each site. Four cores were collected to 0.50 m using a 75 mm diameter core for bulk density (BD).

Table 1. Comparison details and site history for each parent material class. Vegetation types include: native perennial pastures (NPP), introduced perennial pastures (IPP), crop and pine plantation. Soil treatments include: liming and phosphorus (P) application.

Parent material	Comparison	Treatment
Granite	IPP: Limed vs unlimed	2.5t/ha lime broadcast in 2002. IPP sown 1970.
	NPP *vs* <5 yr old IPP *vs* Crop	Crop (wheat): cropped since 2011 (previously IPP). New IPP sown 2011 (previously cropped since 2009 from IPP).
	IPP: High P *vs* Low P	High P site: P management plan since 2005. Low P; nil P.High P and low P application on native pasture
Sedimentary	IPP: Aspect North *vs* South	IPP sown 1989 and pasture improved in 2010
	IPP: <10 yr old *vs* >35 yr old	Old IPP sown in 1974. New IPP sown in 2003.
	>35 yr old IPP *vs* Crop *vs* Pine plantation	IPP sown in 1960. Crop (oats): cropped since 1998. Pine plantation established 2002.
	NPP *vs* <5 yr old IPP *vs* Crop	Crop (wheat): cropped since 2009 (previously NPP). New IPP sown 2010 (previously cropped since 2004 from IPP).
Basalt	NPP *vs* Crop	Crop (barley): first year crop (previously NPP).

Analytical methods
Total Carbon (TC g/100g) was determined on all samples using a LECO (CNS 2000) combustion furnace [4]; Method 6B3). Bulk density (BD) was determined for each core and each depth interval as described by Dane and Topp [5]. Results were calculated as BD in g/cm³ on an oven-dry basis. Results for this paper are reported as C stock in Mg.C.ha calculated by fixed depth C stock (FDCS); FDCS (Mg C ha) = TC (g/100g) x BD (g/cm^3) x depth (cm) x gravel correction factor.

3. Results and Discussion

Parent material significantly influenced the stock of soil C; with basalt- and sedimentary-derived soil having a significantly greater stock of C ($P < 0.05$) in the 0 to 0.50 m compared with granite-derived soil; 77.5 (7.1 sd) *vs* 72.1 (17.2 sd) *vs* 54.9 (9.3 sd) Mg.C.ha respectively. The 0.30 to 0.50 m soil layer contained a considerable proportion of soil C with on average 31 %, 21 % and 24 % measured in the 0.30 to 0.50 m layer for basalt-, sedimentary-and granite-derived soil respectively. This highlights the importance of considering this soil layer when investigating the implications of land management on soil C stocks.

The high P treatment under NPP on granite-derived soil had a greater stock of soil C compared with the low P treatment: 46.8 vs 40.4 Mg.C.ha to 0.50 m (Table 2). Granite-derived soil in the

Monaro region are inherently low in P. Addressing the P requirements of pastures can increase above and below ground biomass and therefore OM supply to the soil [6] and can increase the stability of soil C [7].

The limed treatment under IPP on granite-derived soil had a lower stock of soil C compared with the unlimed site; 62.8 vs 66.7 Mg.C.ha to 0.50 m (Table 2). This is consistent with Chan et al. [6] and may indicate increased microbial decomposition of native soil C when acidic soil constraints are removed by liming.

Interestingly, there was a greater stock of soil C under two of the three younger (<10 and <5 yr old) perennial pastures compared with older pastures (>35 yr old IPP or NPP). The <5 yr old pasture on sedimentary-derived soil which had more soil C (+29.5 Mg.C.ha to 0.50 m) compared with the older pasture was sown in 2010 and had been cropped for the previous 6 years. Therefore this greater stock of soil C may be explained by continued fertiliser (N,P,S) application during both the cropping and pasture phase and the three years of above average rainfall since pasture establishment in 2010.

Two of the cropping sites; one on sedimentary-derived soil and one on basalt-derived soil, had a lower soil C stock compared with their perennial pasture pair; 54.6 vs 60.2 and 72.4 vs 82.5 Mg.C.ha to 0.50 m respectively (Table 2). However, there was no difference between the cropped and pasture site on granite-derived soil and there was a greater stock of soil C under the cropped site compared with the NPP on the remaining sedimentary-derived soil; 60.6 vs 68.0 Mg.C.ha to 0.50 m (Table 2).

Table 2. Carbon stocks (Mg/C/ha) for the 0 to 0.30 and 0 to 0.50 m soil layers. Underlined treatments were used to calculate differences in carbon stock (Mg/C/ha) for a given pair/triplet.

Parent material class and comparison	Treatment	C stock (Mg.C.ha)		Comparison (Mg.C.ha)	
		0 - 0.30 m	0 - 0.50 m	0 - 0.30 m	0 - 0.50 m
Granite derived soil					
NPP: Low P vs High P	NPP Low P	32.92	40.44		
	NPP High P	36.29	46.08	(+) 3.4	(+) 5.6
IPP: Unlimed vs Limed	Unlimed	52.8	66.7		
	Limed	46.9	62.8	(-) 5.9	(-) 3.9
NPP vs <5 yr old IPP vs Crop (wheat)	NPP	41.4	58.5		
	IPP <5 yr old	39.5	51.5	(-) 1.9	(-) 7.0
	Crop	41.4	58.1	0.0	(-) 0.4
Sedimentary derived soil					
IPP: Northern vs Southern Aspect	Northern Aspect	47.3	56.4		
	Southern Aspect	45.7	55.6	(-) 1.6	(-) 0.8
IPP: <10 yr old vs >35 yr old	IPP <10 yr old	75.2	92.7		
	IPP >35 yr old	73.6	96.6	(+) 1.5	(-) 3.9
IPP >35 yr old vs Crop (oats) vs Pines (10 yr old)	IPP >35 yr old	45.2	60.2		
	Crop	42.7	54.6	(-) 2.5	(-) 5.6
	Pines (10 yr old)	67.2	85.7	(+) 22.0	(+) 25.4
NPP vs <5 yr old IPP vs Crop (wheat)	NPP	46.5	60.6		
	IPP <5 yr old	70.3	90.1	(+) 23.8	(+) 29.5
	Crop	52.1	68.0	(+) 5.7	(+) 7.4
Basalt derived soil					
NPP vs Crop (barley)	NPP	61.3	82.5		
	Crop	46.8	72.4	(-) 14.5	(-) 10.1

4. Conclusion

Soil C stocks in the Monaro region varied with parent material and with land use and management. Our baseline information from 19 sites; 8 pairs/triplets, in the Monaro region of southern NSW indicates that nutrient management may increase soil C stocks under perennial pastures. This field survey also highlights the importance of perennial pastures, regardless of pasture age, for the accumulation of C in soil. In this field survey cropping did not decrease soil C stocks at all sites and cropping may therefore offer an opportunity for landholders to diversify their enterprise without compromising soil C in the long-term. However, three of the four cropped sites had only been cropped for less than three years, and these three years coincided with three years of above average annual rainfall in the Monaro region. Furthermore, the vulnerability of C to decomposition (that is, C fractions) was not investigated in this study and it suspected that under different previous rainfall circumstances these results may have been different.

Acknowledgements

The co-operation of the land holders in the Monaro region whose properties were sampled and who provided extensive land management information and continuing interest is gratefully acknowledged. The authors would like to thank Robert Smith (NSW DPI) for field assistance. This project is funded through the Action on the Ground program (Round 1) as part of the Australian Governments Carbon Farming Futures initiative.

References

[1] Chan KY, Oates A, Li GD, Conyers MK, Prangnell RJ, Poile G, Liu DL, Barchia IM 2010 Soil carbon stocks under different pastures and pasture management in the higher rainfall areas of south-eastern Australia. *Soil Res* **48**(1) 7-15.

[2] Peel MC, Finlayson BL, McMahon TA 2007 Updated world map of the Köppen-Geiger climate classification. *Hydrol Earth Syst Sc* **11**, 1633-1644.

[3] Sanderman J, Baldock J, Hawke B, MacDonald L, Massis-Puccini A, Szarvas S 2011 National Soil Carbon Research Programme: Field and Laboratory Methodologies. CSIRO, Adelaide.

[4] Rayment GE, Higginson FR 1992 'Australian Laboratory Handbook of Soil and Water chemical methods.' (Inkata Press: Melbourne)

[5] Dane JH, Topp CG (Eds) 2002 'Methods of soil analysis: Part 4 Physical Methods.' Agronomy ; no. 9 (Soil Science Society of America, Inc: Madison, Wis. USA)

[6] Chan KY, Conyers MK, Li GD, Helyar KR, Poile G, Oates A, Barchia IM 2011 Soil carbon dynamics under different cropping and pasture management in temperate Australia: Results of three long-term experiments. *Soil Res* **49**(4) 320-28.

[7] Kirkby CA, Kirkegaard JA, Richardson AE, Wade LJ, Blanchard C, Batten G 2011 Stable soil organic matter: A comparison of C:N:P:S ratios in Australian and other world soils. *Geoderma* **163**(3–4), 197-208.

IOP Conf. Series: Earth and Environmental Science **25** (2015) 012019 doi:10.1088/1755-1315/25/1/012019

Can soil change be assessed for the Victorian dairy industry?

Sharon R. Aarons[1], Douglas Crawford[1], Mark Imhof[2], Cameron Gourley[1]

[1] Agriculture Research Division, Department of Economic Development, Jobs, Transport and Resources, 1301 Hazeldean Road, Ellinbank, 3821, Australia
[2] Agriculture Research Division, Department of Economic Development, Jobs, Transport and Resources, 32 Lincoln Square North, Carlton, 3053, Australia

E-mail: sharon.aarons@ecodev.vic.gov.au

Abstract. Meeting the increased demand for dairy products will require careful management of soils to minimise land degradation and sustain increased production. Key to providing farmers with the tools to manage their soils sustainably, is firstly understanding the soil types currently managed by dairy farmers, and secondly quantifying changes in soil properties in response to management. The Victorian Land Use Information System was interrogated to identify dairy land parcels and these data overlaid on soil survey information to identify the dominant soil orders managed by dairy farmers in the three dairy regions of Victoria. Of the approximately 590,000 hectares of dairy land identified across the state, Sodosols (33%), Chromosols (20%), Dermosols (16%), and Vertosols (11%) are the major soil Orders represented, although the dominant soil Orders vary for each region. Legacy data from research and extension activities undertaken between 1995 and 2010 were collated to understand regional differences in dairy soil properties. All soil properties were significantly and positively skewed with higher median pH, EC and available K in northern Victorian soils. Further analysis compared the 1995 to 2010 data with data from samples analysed by the government analytical laboratory between 1973 and 1980 to assess any differences over 38 years. The older soil chemical data were also positively skewed but had lower median soil pH, Olsen P and available K, consistent with the greater use of inputs by the industry in more recent years.

1. Introduction

The Australian dairy industry like other agricultural sectors is responding to increasing world demand for food [1], especially as the economies of countries with traditionally lower gross domestic product improve and diets include more animal protein. While dairy farms have intensified production [2] a major competitive advantage of the industry remains the relatively low cost of producing permanent pasture. A number of farm management practices have the potential to degrade soil condition and therefore reduce plant growth and compromise animal health. These include how nutrients and forage are applied and utilised, and nutrient redistribution through regular on-farm movement of dairy cows. For example, the dairy industry has increased its use of fertiliser, feed and effluent nutrients over the past 20 years with average use of N, P, K and S fertiliser of 96, 30, 63, and 26 kg ha^{-1} respectively, but up to 424, 82, 177, and 85 kg ha^{-1} respectively applied on some farms [3]. Cows can deform soils due to the size and weight of the animal [4], and the nutrient content of their excreta will affect soil chemical and biological properties [5, 6]. To help ensure that dairy production systems are sustainable, managers require knowledge of a) soil types and properties b) the condition of the soils and c) the impact of practices. This paper describes the approach taken to meet the first two requirements and discusses the implication of the results.

2. Material and Methods

2.1 Victorian dairy soil types

The Victorian Land Use Information System (VLUIS) database which provides spatial coverage of land use across Victoria [7] was used to identify 'dairy' parcels within the 'pasture/grassland' land cover classification. The dairy parcels were overlaid with available regional and sub-regional soil survey mapping. This mapping varied from detailed soil type mapping (e.g. in irrigation regions) to broader scale soil/landform (1:100 K scale) or land system (1:250 K scale) mapping. The dairy soils identified were grouped according to their dominant soil Orders based on the Australian Soil Classification [8].

2.2 Victorian dairy soil condition

Legacy soil data including unpublished data from Victorian Government researchers and extension staff were collated to provide estimates of dairy soil condition. These data consisted of 4814 records collected from 16 experimental and extension activities undertaken on 297 dairy farms around the state between 1995 and 2010 (Table 1). Additional data from samples collected between 1973 and 1980 were sourced from the Victorian Government soil analytical laboratory database. These data (3738 records), derived from samples collected by Government extension staff from dairy paddocks, were compared with the more recent (1995 to 2010) dataset. Samples had been collected using recommended procedures and accredited analytical methodologies were used.

Table 1. Experiments from which legacy soil data were sourced.

Data ID	Dates	Farms sampled	Regions$^{\Phi}$	Reference
A4N	2008	17	Gi, swV, NV	[9]
MDF	2007-2010	1	Gi	Lane pers comm.
AS	2007-2008	28	swV	Standish pers comm.
HDLN	2006-2007	24	swV	[10]
MRF	2005	1	Gi	Neilsen et al. pers comm
NMP_FSV	2005-2008	142	Gi	Kelsall pers comm.
HBI	2004-2006	2	Gi	Gourley et al. pers comm.
TAR	2004	4	Gi	Melland et al. pers comm.
BBF	2004	2	Gi	Gourley et al. pers comm.
DSAT	2003-2004	30	Gi, swV, NV	Neilsen et al. pers comm
GDRP-P	2002	1	Gi	[11]
WFNP	1999, 2002	3	Gi	Gourley et al. pers comm
NWH	1999	4	Gi	White pers comm
CSOP	1998	36	Gi, swV, NV	[12]
SQ	1997-1998	1	Gi	[13]
PfDF	1995-2001	10	Gi	[14], DEDJTR Research Farmlets

$^{\Phi}$Victorian dairy regions: Gi (Gippsland), swV (south-west Victoria), NV (northern Victoria)

2.3 Sample collection and analysis

The soil samples consisted of a minimum of 30 soil cores (2.5 cm diameter x 10 cm depth) collected either randomly throughout a paddock or systematically along a transect considered representative of the paddock. Dried (40°C), and sieved (<2 mm) samples were analysed for pH (in water or 0.01 $CaCl_2$) and electrical conductivity (EC) at a 1:5 soil:solution ratio. Bicarbonate extractable Olsen P, available K (either by bicarbonate extraction – Colwell K, or hydrochloric acid - Skene K), and S (KCl_{40}-S) were analysed. Details of the methods are provided in Rayment and Lyons [15].

2.4 Statistical analysis
The soil order data were summarised to give percentages of each soil type. Legacy data were analysed to quantify chemical properties of soils under dairy management. Summary statistical analysis was initially carried out using Genstat 14.1 (VSN International) to calculate the mean, median, minimum, maximum, standard deviation and coefficient of variation. The data were also analysed to assess regional (Gi – Gippsland, NV – northern Victoria, swV – south- west Victoria) soil differences using REML analysis, and to compare past (1973 to 1980) and more recent (1995 to 2010) soil properties.

3. Results and Discussion
The Victorian dairy industry is approximately evenly distributed in three regions; northern Victoria, Gippsland and south- west Victoria, based on dairy farm post code data (Figure 1, Table 2, [16]). The soil types in each region are likely to influence soil properties and therefore recommended soil management practices.

Table 2. Numbers (and percentages) of Victorian dairy farms (2011/2012), legacy research and extension (1995-2010) and analytical (1973-1980) data for Gippsland (Gi), northern Victoria (NV) and south-west Victoria (swV) regions.

Region	Dairy farms	Research and extension records	Government analytical lab records
Gi	1575 (34)	2490 (52)	1382 (37)
NV	1551 (34)	392 (8)	840 (22)
swV	1454 (32)	1932 (40)	1518 (41)

3.1 Victorian dairy soil types
A total of 590,000 hectares of land was estimated, using the VLUIS, to be managed by dairy farmers in Victoria. Sodosols (33%), Chromosols (20%), Dermosols (16%) and Vertosols (11%) accounted for 80% of Victorian soils, with the remaining eight soil Orders each comprising less than five percent (data not presented). However, the proportion of major soil Orders varies in each region (Figure 1). For example, Sodosols, the dominant soils in the NV (61%) and Gippsland (19%) regions, only comprise an estimated 9% of the soils managed by dairy farmers in south- west Victoria. The dominant soil Orders reflect characteristics such as sodicity, acidity and degree of texture contrast that will affect their response to typical farm management practices as well as influence how these soils should best be managed to minimise their chemical and physical constraints to pasture and fodder production and grazing management. In addition to regional differences in dominant soil orders, the proportion of Sub-orders also vary within each region. For example, Black Dermosols are more common in the eastern part of Gippsland (alluvial plains) compared with Brown Dermosols to the west in the rolling hills. These differences also influence soil characteristics such as depth of topsoil and friability.

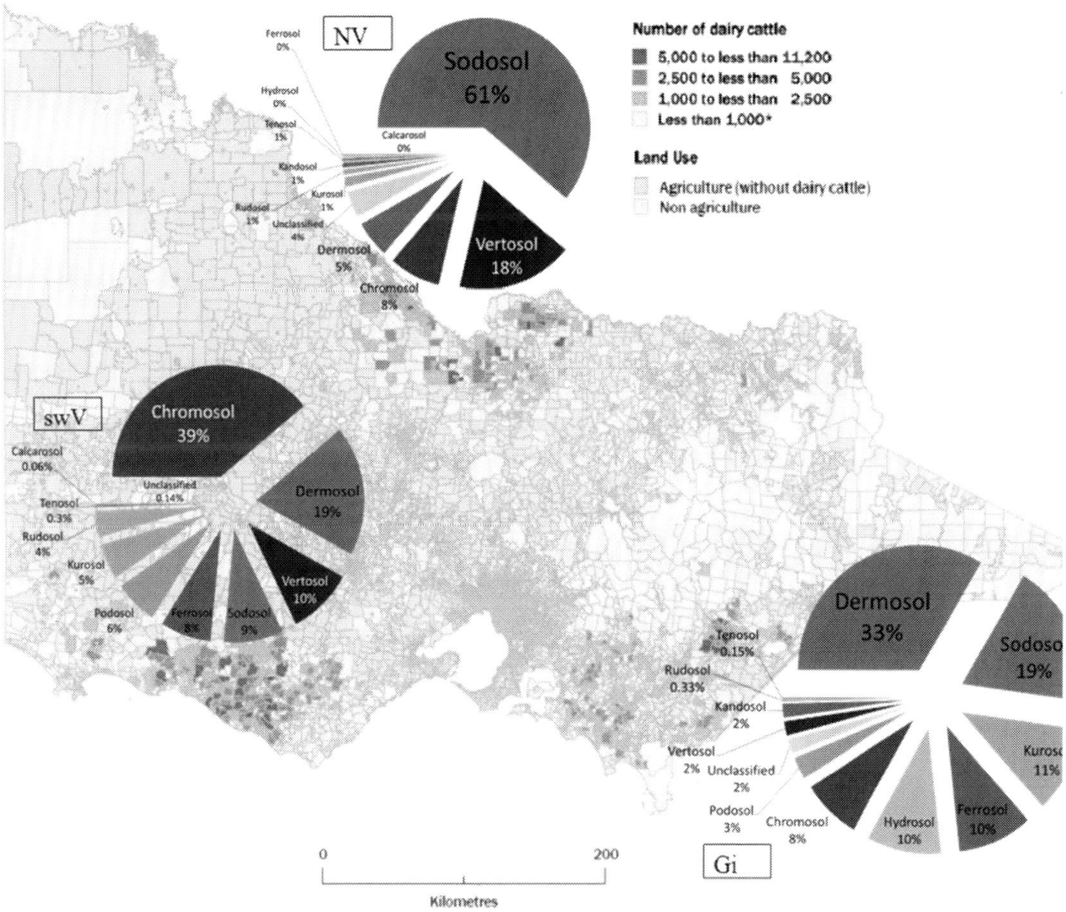

Figure 1. Locations of the Victorian dairy industry (number of dairy cattle; [17]) and dominant soil Orders in the northern Victoria (NV), Gippsland (Gi), and south-west Victoria (swV) regions (derived from the VLUIS and Victorian soil survey data).

3.2 *Victorian dairy soil change*

Legacy research and extension data were available from all regions with the majority (52%) from dairy farms in Gippsland and only 8% from northern Victoria (Table 2); this distribution reflecting the location of the research and extension groups and programs as well as the relative priority placed on irrigation rather than nutrient management in northern Victoria. Soil chemical properties were significantly skewed with higher median soil pH, EC and available K in northern Victoria; most likely influenced by the dominant soil orders in that region (Table 3, Figure 2a). Significant ($P<0.04$) region and farm differences and non-significant paddock differences were observed for soil properties.

Table 3. Summary statistics of 4814 records of legacy chemical data from soil samples collected on dairy farms in all three regions in Victoria between 1995 and 2010.

Soil property	No. of samples	Minimum	Mean	Median	Maximum	Std Dev	CV (%)	Skew
pH (H_2O)	4156	4.5	5.7	5.6	8.4	0.50	9	0.95
pH ($CaCl_2$)	3576	3.8	5.0	5.0	8.1	0.54	11	1.13
EC (dS m^{-1})	3615	0.02	0.18	0.14	7.31	0.218	120	17.995
Olsen P (mg kg^{-1})	4814	3	38	33	711	27.6	74	8.5
Avail K (mg kg^{-1})	4814	39	319	270	11000	334.5	105	17.8
S_{KCL-40} (mg kg^{-1})	4397	2	19	15	810	23.6	122	16.2

The 1973 to 1980 legacy data were more uniformly distributed around the state in line with the distribution of dairy farms (Table 2). As with the 1995 to 2010 data, soil chemical properties were also positively and significantly skewed (data not presented). However, median soil pH, K, and Olsen P from the earlier samples were lower than that from 1995 to 2010. These differences are mostly likely due to the greater use of inputs by the industry recently (Figure 2b,c, [3]).

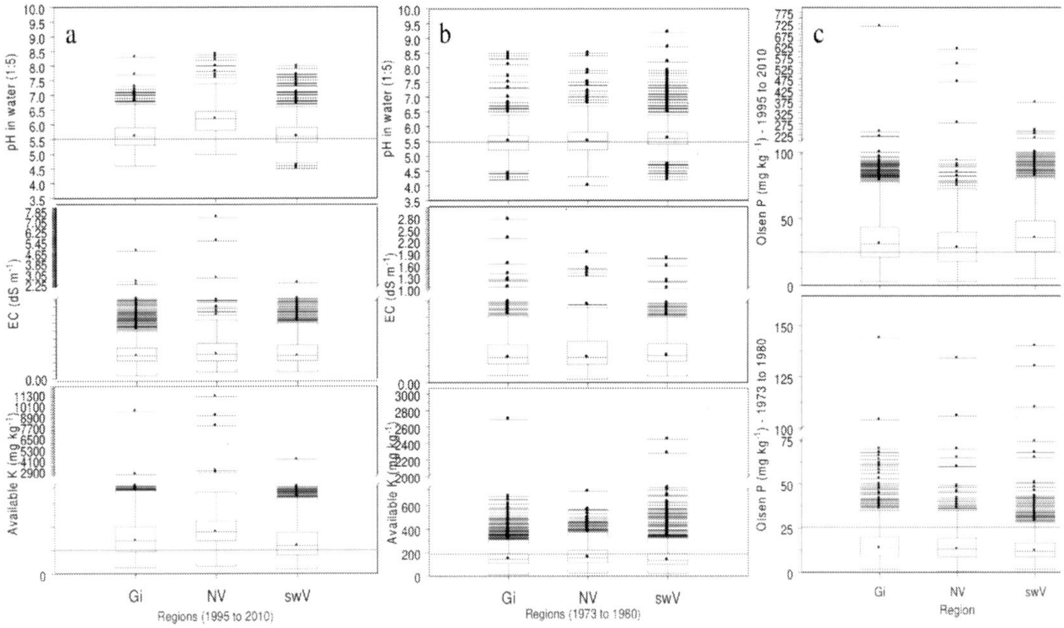

Figure 2. Soil pH and EC (dS m^{-1}) (1:5 soil: water) and available K (mg kg^{-1}) for soils sampled on Victorian dairy farms between a) 1995 and 2010, b) 1973 and 1980, and c) Olsen P (mg kg^{-1}) data from the same two time periods for the Gippsland (Gi), northern Victoria (NV) and south-west Victoria (swV) regions. Horizontal red lines indicate agronomic optimum levels.

4. Conclusions

The dominant soil orders identified for each dairy region can be used as a basis for developing region-specific soil and farm management guidelines and programs for the industry. Analysis of legacy data has been used for understanding soil change more broadly in Victoria [17], and in this paper industry specific legacy data were investigated to understand regional and temporal differences in soil properties. While legacy data can be used to inform priority farm management and policy decisions and to strategically target soil monitoring resources, data analysis and interpretation needs to be mindful of sample collection and analytical methods and fit for purpose sampling.

References

[1] FAO 2009 How to feed the world in 2050.
[2] Dairy Australia 2010 Australian Dairy Industry In Focus 2010.
[3] Gourley CJP, Dougherty WJ, Weaver DM, Aarons SR, Awty IM, Gibson DM, Hannah MC, Smith AP, Peverill KI 2012 Farm-scale nitrogen, phosphorus, potassium and sulfur balances and use efficiencies on Australian dairy farms. *Anim Prod Sci* **52** 929-44.
[4] Bilotta GS, Brazier RE, Haygarth PM 2007 The impacts of grazing animals on the quality of soils, vegetation, and surface waters in intensively managed grasslands. *Adv Agron* **94** 237-80.
[5] Williams P, Haynes RJ 1995 Effect of sheep, deer and cattle dung on herbage production and soil nutrient content. *Grass Forage Sci* **50**263–71.
[6] Aarons SR, O'Connor C, Hosseini HM, and Gourley CJP 2009 Dung pads increase pasture production, soil nutrients and microbial biomass carbon in grazed dairy systems. *Nutr Cycl Agroecosys* **84**81–92.
[7] Morse-McNabb E 2011 The Victorian Land Use Information System (VLUIS): A new method for creating land use data for Victoria, Australia. Proceedings of the Surveying and Spatial Sciences Biennial Conference New Zealand
[8] Isbell R 2002 The Australian Soil Classification. CSIRO Publishing, Melbourne.
[9] Gourley C, Dougherty W Aarons S, Hannah M 2010 Accounting for nutrients on Australian dairy farms, Final Report to Dairy Australia, DAV 12307, June 2010, Department of Primary Industries, Ellinbank, Victoria.
[10] Greenwood K, Rees D, Davey M, Brown A 2008 HDLN Soil and water dairy action program, Final Report to Dairy Australia, DAV 12222, June 2008, Department of Primary Industries, Kyabram and Werribee, Victoria.
[11] Aarons S 2006 Gippsland dairy riparian project: Environmental monitoring, Final Report to Dairy Australia, DAV11278, Department of Primary Industries, Ellinbank, Victoria.
[12] Aarons S 2001 Cycling of organic phosphorus in grazed dairy systems, Final Report to DRDC, DAV392, Agriculture Victoria, Department of Natural Resources and Environment, Ellinbank, Victoria.
[13] Crawford A, Hirth J, Gourley C 1999 Maintaining soil quality of grazed pastures, Final Report, SAS project 02249/12134, September 1999, Agriculture Victoria, Department of Natural Resources and Environment, Ellinbank and Rutherglen, Victoria.
[14] Gourley C 2001 Phosphorus for dairy farms 1995-2001, Final Report to DRDC, DAV318, October 2001, Agriculture Victoria, Department of Natural Resources and Environment, Ellinbank, Victoria.
[15] Rayment GE, Lyons D 2010 Soil Chemical Methods - Australasia
[16] DEDJTR 2015 Dairy-Industry-Profile_December-2014-Update; last accessed February 2015
[17] Marchant BP, Crawford DM, Robinson NJ 2014 What can legacy datasets tell us about soil quality? Soil Change Matters Symposium, Bendigo, Australia.

Delivering solutions to questions regarding soil change—examples from USDA and the National Cooperative Soil Survey

Maxine J Levin[1], Susan Andrews[2], Michael Robotham[3], Joel Brown[4], Lenore Vasilas[5] and David Hoover[6]

[1, 5] USDA Natural Resources Conservation Service, 5601 Sunnyside Avenue, Beltsville, MD, USA

[2, 3, 6] USDA Natural Resources Conservation Service, 100 Centennial Mall North, Room 152, Lincoln, NE, USA

[4] USDA Natural Resources Conservation Service, Jornada Experimental Range, MSC 3JER, New Mexico State University, Box 30003, Las Cruces, NM, USA

E-mail: maxine.levin@wdc.usda.gov

Abstract. The US National Cooperative Soil Survey (NCSS) has been investigating Soil Change within the Interpretations Conference Committees and within the USDA-Natural Resources Conservation Service (NRCS), Soil Science Division since the early 1990's. Historically, the National Cooperative Soil Survey worked on building a national map that would deliver information on land use and soil management. Soil Mapping, Soil Classification and Soil Taxonomy focused on the static qualities of the soil profile, attempting to make estimates and predictions of soil groupings based on soil characteristics that were stable beyond a 5 to 20 year cycle of use and management and potential anthropogenic change. The National cooperative Soil Survey continues to seek new ways to interpret soils and to make that information more easily accessible. Practice-specific soil interpretations are being explored to support NRCS conservation planning and practice implementation. Other areas of focus include the development of real-time interpretation systems that allow incorporation of site-specific information and interpretation systems that will allow users to incorporate other spatially-referenced data sets including climate and land use to develop accurate and site-specific interpretive information products.

1. Introduction

The US National Cooperative Soil Survey has been investigating Soil Change within the Interpretations Conference Committees and within the USDA-Natural Resources Conservation Service (NRCS), Soil Science Division since the early 1990's. Historically, the US Soil Survey worked on building a national map that would deliver information on land use and soil management. In the US, all these assessments of the potential uses of soils, from agricultural production to engineering properties, have become consolidated under the overarching category of Soil Interpretations. Soil Mapping, Soil Classification and Soil Taxonomy focused on the static qualities of the soil profile, attempting to make estimates and predictions of soil groupings based on soil characteristics that were stable beyond a 5 to 20 year cycle of use and management and potential anthropogenic change.

From the early years of the soil survey through the development of computer databases in the 1970s, soil interpretations were based on written guides that were used by the soil surveyors to develop the interpretation tables contained in soil survey manuscripts published by the National Cooperative Soil Survey. Tables contained use and management interpretations by map unit or component. Interpretive results for a tract of land could be determined by cross referencing the hard copy soil map and the interpretive table(s). Examples of common interpretations provided in soils surveys include: crop, forage and range suitability groups, and use limitations for recreation, building site development, and engineering uses. The question of soil change focussing on the potential for erosion, salinity and sodicity, soil properties of wetlands, drainage phases and soil contamination has been the realm of soil interpretations and tying the map unit and individual soil characteristics to suitability or vulnerability indexes in separate tables or maps.

In the mid 1980's through 2000 the National Cooperative Soil Survey started to explore how we might map the soil survey information to display more information by soil function and the capacity to recover from various anthropogenic forces. Farm and rangeland areas with drastic soil disturbances from native landscape were being mapped and classified within US Soil Taxonomy protocols but did not describe soil changes enough to categorize for interpretations or document changes from original material to inform management practices. The National Cooperative Soil Survey also started mapping urban areas where all the soils in the survey area had anthropogenic characteristics that needed to be described, classified and interpreted for management purposes. Soil change issues within policy forced us to build new systems of classification and description.

2. Delivering solutions with integration of policy and science

A prime example of the integration of policy and science in the National Cooperative Soil Survey was hydric soil indicators for wetland delineation as part of the Food Security Act and Clean Water Act. Disturbed functioning wetlands and drained systems were being considered for regulation by the Environmental Protection Agency (EPA) and Army Corp of Engineers as "Waters of the US." Functioning wetlands needed to be delineated and documented. Wetland classes needed to be interpreted from existing soil maps and historical aerial photography. There was extensive analysis of a five-year history of historical compliance imagery. Policy required the identification of both functional hydric soils for jurisdictional wetland determinations and drained or non-functional hydric soils for wetland restoration practices for programs such as the wetland reserve program. The National Cooperative Soil Survey partnership worked with other agencies and scientists, building from the established soil and flooding hazard maps, to create " high potential" wetland maps to inform the public.

Hydric soil lists were established by National Cooperative Soil Survey teams and published for public record but were not used as definitive information for wetland delineation. Onsite examination of hydric soil indicators to confirm active function of wetlands for policy management and regulatory statutes became the gold standard to delineate wetlands and document wet conditions. Traditional Soil Surveys mapped hydric soils based on their classification but on-site determinations using the field indicators are now needed to identify those areas that continue to function as a hydric soil (Figure 1) [1] from those that (through changes in soil and hydrologic characteristics) are no longer functional hydric soils.

IOP Conf. Series: Earth and Environmental Science **25** (2015) 012020 doi:10.1088/1755-1315/25/1/012020

Figure 1. Online and published guide-- Field Indicators of Hydric Soils in the United States. A Guide for Identifying and Delineating Hydric soils, Version 7.0, 2010. National Interagency committee for Hydric Soils collaboratively directed compilation of indicators to reflect wetland soil change and address national policy needs to identify active wetlands by soil characteristics.

In addition to the example of hydric soil policy implementation, there have been other programs (highly erodible land with wind or water soil erosion potential; salinity vulnerability and reclamation; and subsidence vulnerability) that have been providing guidance for conservation protection and yet allowed some flexibility for the science community to continue research for refinement and improved definition of terms and outcomes. The science community can learn from the example of hydric soil indicators development in the late 1980's through early 2000's on how to approach the relationships between policy formulation, science and regulation. There needs to be a balance between continual collection and validation of new information on soil change and the push of climate change adaptation policy progress.

The National Cooperative Soil Survey continues to identify ways to improve existing products and develop new products. The US Soil Survey has been transitioning from data collection and summary for empirical modeling and estimated values to point data summary for process modeling which could produce values and outcomes useful to climate change adaptation policy. The idea that specific measured or estimated point data can be spatially extrapolated through process modelling to predict outcomes of climate change is only now being explored in government sponsored environmental monitoring communities. As we complete initial US Soil Survey inventory projects, we are allowing for more resources to do on-site characterization specifically for process modelling of conservation practices on soil function, nitrate leaching, soil erosion and potential commodity crop yields. Conservation Effects Assessment Project (CEAP) is an interagency effort to quantify the environmental effects of conservation practices and programs and develop the science base for managing the agricultural landscape for environmental quality. Using the US Soil Survey as a base data resource for spatial analysis and additional project sampling for validation, assessments in CEAP are carried out through process modelling at national, regional and watershed scales on cropland, grazing lands, wetlands and for wildlife. The three principal components of CEAP—the national assessments, the watershed assessment studies, and the bibliographies and literature reviews—contribute to building the science base for conservation. That process includes research, modeling, assessment, monitoring and data collection, outreach, and extension education. CEAP findings are being used to guide USDA conservation policy and program development and help conservationists, farmers and ranchers make more informed conservation decisions [1].

IOP Conf. Series: Earth and Environmental Science **25** (2015) 012020 doi:10.1088/1755-1315/25/1/012020

Next steps for further work will also be an increased focus on soil monitoring through the US National Resource Inventory (NRI) and the Ecological Site Inventory [1]. Soil survey mapping data technology and modeling teams are working together to identify soil change and vulnerable landscapes. In addition, the modeling effort through collection of long term monitoring with NRI tied to the spatial soil data base has created an opportunity to analyze conservation effects of management to land resources. Capacity for natural resource assessment and analysis has been developed through Conservation Effects Assessment Program (CEAP) by integrating investments such as the National Resources Inventory (NRI), Soil Survey base data through Gridded Soil Survey Geographic (gSSURGO) geospatial databases (Figure 2) [2], conservation practice implementation data, and partner monitoring data—with powerful and improved analytical models and methods [3]. In the last 5 years, the National Cooperative Soil Survey soil change dialog and the USDA NRCS Soil Science and Resource Assessment Deputy Area have aligned priorities to build on research progress, monitoring programs and modeling capacity. Modeling teams within the US National Cooperative Soil Survey are now forming at the regional level to build capacity for further development and integration.

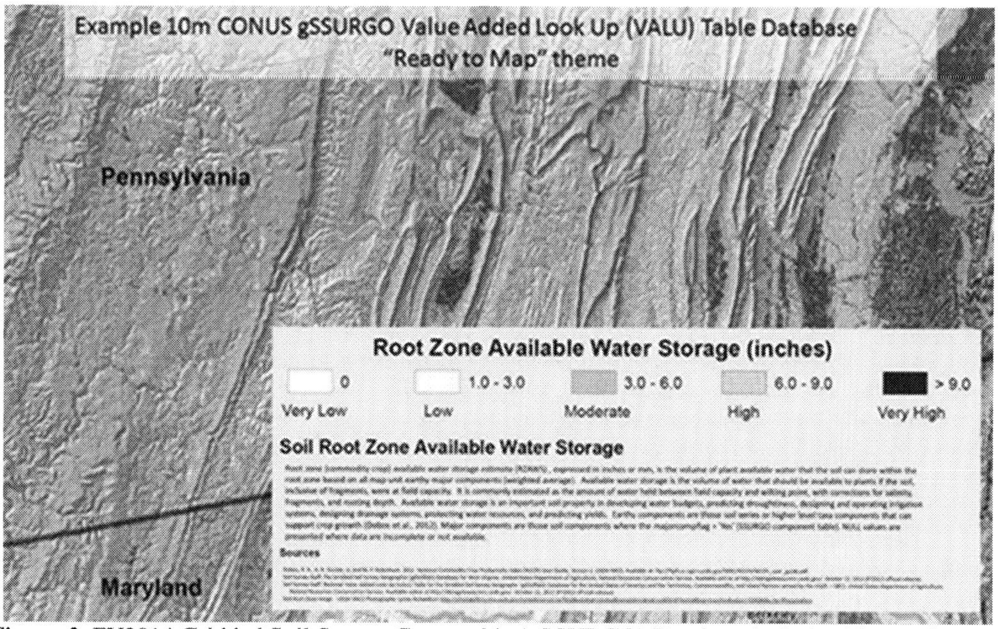

Figure 2. FY2014 Gridded Soil Survey Geographic (gSSURGO) Database is derived from a January 15th, 2014 snapshot of the Soil Data Mart database. These new data are available in both state-wide tiles and the Conterminous U.S. (CONUS) [2].

The USDA-Natural Resources Conservation Service (NRCS), National Soil Survey Center, Lincoln, Nebraska, leads a group that coordinates studies in Soil Quality and Ecosystem Dynamics. The group focuses on the development of useful tools for measurement and applications for policy related to Soil Change, Resistance, Resilience, State and Transition Models (STMs) and Dynamic Soil Properties (DSPs). The National Cooperative Soil Survey started committees in 2010 which have been consolidating research and evaluating inventory techniques and data to push forward a practical and scientifically robust approach to mapping and applying soil change properties to spatial soils data. This entails teams addressing topics such as onsite assessment indicators, sampling protocols, and developing data models for data storage within the National Soil Information System (NASIS). The committees and Dynamic Soil Property Team at the National Soil Survey Center in Lincoln NE have developed assessment protocols and a 3 year plan for sampling and research starting in 2014.

IOP Conf. Series: Earth and Environmental Science **25** (2015) 012020 doi:10.1088/1755-1315/25/1/012020

Priorities for the committees are documented and are summarized as follows: (http://www.nrcs.usda.gov/wps/portal/nrcs/detail/soils/survey/partnership/?cid=nrcs142p2_053534#ecosystem) [1]

1. Identify criteria for a minimum datasets for DSP and ecological sites (ESs), especially for new land types.
2. Develop and support a database that contains the correct data elements to relate vegetation composition to DSPs and vice versa. Allow data retrieval and queries.
3. Develop partnerships to share data in a format that is compatible, linked to soils, and at a useable scale. Review current pathways for NRCS agency and other users to access completed ESDs.
4. Use data analysis tools and skills to analyze data for ESD and DSP interpretations as drivers of ecosystem change. Support standardization of properties and protocols for ESs and DSP data collection
5. Report criteria that are being used to differentiate sites and states for various land types
6. Explore the adaptation of ESs for use on croplands
7. Discuss how dynamic soil properties can be effectively integrated into all ESs

The National Ecological Site Description Initiative, (now led by Joel Brown, USDA-NRCS and Agricultural Research Service (ARS) at Jornada Experimental Range in Las Cruces, New Mexico), is providing technical support to teams throughout the US to develop Ecological Site Descriptions that describe impacts of management or climate change on the landscape. The ARS Jornada has been partnering for many years through the National Cooperative Soil Survey to build a framework for the National Ecological Site Inventory. Ecological sites are conceptual divisions of the landscape based on soil map unit components as defined and mapped by the National Cooperative Soil Survey. Recurring soil, landform, geological, and climate characteristics are grouped together based on their dynamic responses to disturbance, whether natural or anthropogenic. The differing soil: plant communities determine ecosystem services that may be derived from land and management interactions. Each ecological site may exhibit multiple stable plant communities (states) and dynamic patterns of change (transitions). State and Transition models (*sensu* [4]) are graphic and textual models of changes in plant communities and associated dynamic soil properties that can occur on an ecological site. The causes of change, the constraints to reversibility of the change, and the management interventions needed to prevent or initiate change are described qualitatively. Together, ecological sites and state and transition models provide spatially-explicit information about multiple potential soil:plant communities and the descriptions of soil change for landowners and agencies to help guide land management decisions (Figure 3) [1,5]. A good overview of Ecological Site history, concepts, applications and protocols is in *Rangelands* December, 2010 [6].

IOP Conf. Series: Earth and Environmental Science **25** (2015) 012020 doi:10.1088/1755-1315/25/1/012020

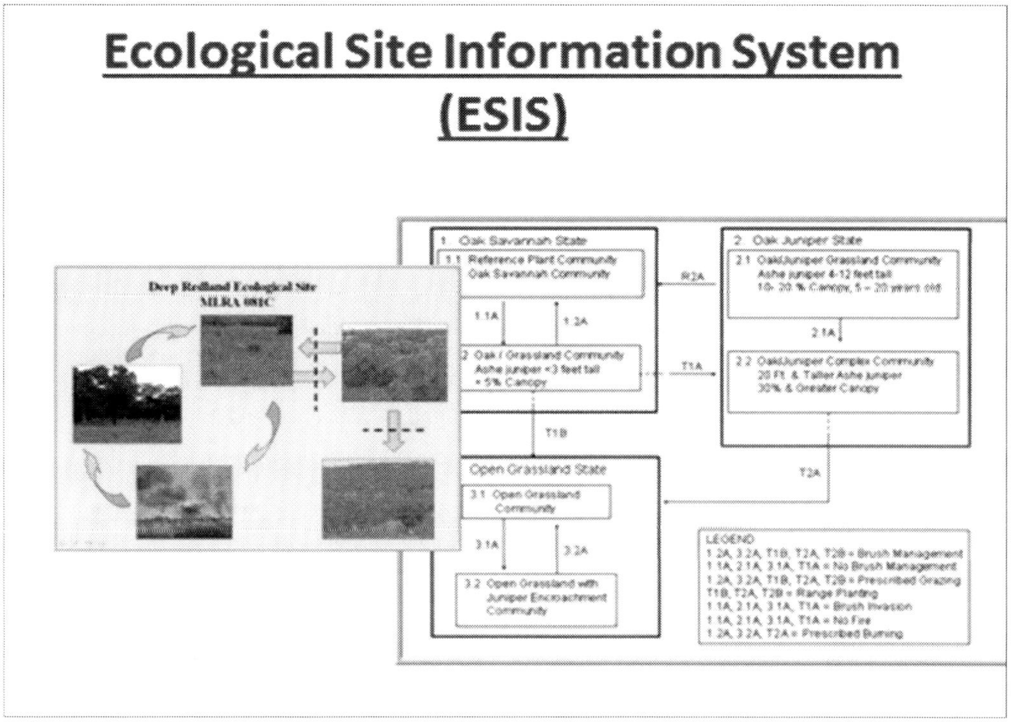

Figure 3. Ecological Site Information (ESIS) is the NRCS repository for ecological site descriptions and for information associated with the collection of forestland and rangeland plot data. ESIS is organized into two applications and associated databases; the Ecological Site Description (ESD) application and the Ecological Site Inventory (ESI) application [1].

There are two National Handbooks on the mapping and documentation of Ecological Sites that were released in 2013 and early 2014 [1]. The first, the Interagency Ecological Site Handbook for Rangelands, was released in early 2013 by the USDA Natural Resources Conservation Service, Bureau of Land Management, and the US Forest Service. The publication of the handbook in 2013 effectively links a unified set ecological concepts and methodological approaches to the land management activities in rangelands for the three major federal agencies. This was a response, in part, to direction from Congress in the Department of the Interior and Related Agencies Appropriations Act of 2002. In that Appropriations Act, Congress expected the Secretary of Agriculture and the Secretary of the Interior to prepare a coordinated plan and budget that would identify the cost of completing standardized soil surveys and ecological classification on all rangeland for use at local management levels. Following the release of the first Rangeland Ecological Site Handbook, the National committees pressed on to finish the editing and vetting of the 2nd more comprehensive Handbook for all of the US, the National Ecological Site Handbook. It has now been released for immediate adoption in early 2014. The Handbooks describe most accepted and vetted definitions as they now stand of Ecological Site Inventory terms, approaches to data collection and standardized analysis nationwide [3].

3. Integration of policy and science with Soil Survey Interpretations

By building on its long history and record of success in interpreting soils information to meet user needs and incorporating new ideas and technology, the Soil Science Division, the NRCS and the National Cooperative Soil Survey partnership are well positioned to continue to deliver interpretive information that meets customer needs now and on into the future. Ongoing activities within the

National Cooperative Soil Survey include improving the consistency of the soils data that underpins interpretations across political boundaries and developing a "minimum data set" of commonly used interpretations to include in digital soils data [3]. Soil Scientists in the field are working with local cooperators to assess and improve existing interpretations, particularly in the context of soil change and climate change adaptation. The National Cooperative Soil Survey continues to seek new ways to interpret soils and to make that information more easily accessible. As an example, practice-specific soil interpretations are being explored to support NRCS conservation planning and practice implementation. Other areas of focus include the development of real-time interpretation systems that allow incorporation of site-specific information and interpretation systems that will allow users to incorporate other spatially-referenced data sets including climate and land use to develop accurate and site-specific interpretive information products. On the distribution side, efforts are underway to look at how interpretive information can be effectively delivered through other avenues including via smart phones and tablets. This will lead to more effective sampling, data collection and interpretation of soil information to support monitoring and soil change attributes.

Soil Change, as a scientific directive and focus, is a new step in the history of the National Cooperative Soil Survey in the United States. Redirecting resources of the soil survey community to linkages with soil characterization, pedo-transfer functions and modeling of processes for yield and erosion outcomes has been useful to better understand soil change. It has put us on a path to acknowledge soil change as a soil property that can be characterized and potentially mapped. Costs of time and money are still involved for monitoring but the framework of state and transition models offers some relief from multi-year projects involving the whole country or a region. Exchange of "space for time" in dynamic soil properties (DSP) monitoring projects with state and transition models [5] has opened up the possibility of predicting and spatially illustrating use and management effects on soil properties. Though extensive "space for time" projects are still several years away in the planning stages for many regions of the United States, the potential payback for a quick turnaround of science-based information in soil change is promising. Spatially armed with models and maps, a farmer, land user or policy maker could be able to make better decisions in the face of food security issues and climate change with these tools in hand.

References

[1] USDA-Natural Resources website http://www.nrcs.usda.gov/ (for National Resource Inventory, CEAP, National Cooperative Soil Survey, hydric soil indicators, Ecological Site Inventory and other topics addressed in this presentation)

[2] Soil Survey Staff, Natural Resources Conservation Service, United States Department of Agriculture. Soil Survey Geographic (SSURGO) Database for [Survey Area, State]. Available online at http://websoilsurvey.nrcs.usda.gov/. Accessed [2014].

[3] Soil Survey Staff, Natural Resources Conservation Service 2014 National soil survey handbook. Title 430-VI. U.S. Government Printing Office, Washington D.C. Sec 602.

[4] Westoby M, Walker and I Noy-Meir 1989 Opportunistic management for rangelands not at equilibrium. *J Range Manage* **42** 266-274

[5] Bestelmeyer BT, Goolsby DP and Archer S R USDA-ARS, Jornada Experimental Range and Jornada Basin LTER and New Mexico State University, Las Cruces, NM; and School of Natural Resources and the Environment, University of Arizona, Tucson, "USA Spatial perspectives in state-and-transition models: a missing link to land management?", *J Appl Ecol* **48** 746–57

[6] Rangelands 2010 Ecological Sites. *Rangelands* **32** (6) 2-64

IOP Conf. Series: Earth and Environmental Science **25** (2015) 012021 doi:10.1088/1755-1315/25/1/012021

Soil Archives: supporting Research into Soil Changes

Linda Karssies[1], **Peter Wilson**
[1] CSIRO Land and Water, GPO Box 1666, ACT, 2601, Australia

E-mail: linda.karssies@csiro.au

Abstract. Soil archives provide valuable support to soil research by making soil specimens and associated data available, reducing the need for labour intensive and expensive fieldwork.

1. International Soil Archives

The usefulness of soil archives has been well demonstrated. Work based on the Rothamsted collection in the United Kingdom provided evidence of steady rates of increase in concentrations of dioxins during the last century [1]. The sample archive of the equally well known Hubbard Brook Experimental Forest in the U.S. led to the discovery of the link between the use of fossil fuels and increased acidification of rain and snow in North America [2]. Soil archives are also the Rosetta stone of terrestrial carbon cycling because atmospheric CO_2 is spiked by radiocarbon from aboveground nuclear weapons testing and offers a tracer for carbon in soil.

2. Australia's National Soil Archive

The CSIRO National Soil Archive was established in 2003 to consolidate several smaller CSIRO collections [3]. The soil specimens from these original collections had been collected for a wide range of research projects and differ in the amount of soil sampled, sampling depth and data collected in the field and in the chemistry laboratories. Since 2003, many state agencies archives and individual projects have contributed soil specimens and data, helping to build a truly National Soil Archive facility. Currently the collection holds 70,000 soil specimens from 9,000 profiles (Figure 1).

The National Soil Archive has protocols both for sample submission and sample use. The National Soil Archive's mission is to provide facilities and protocols for conserving the long-term scientific value of soil specimens and associated soil data. Archived specimens and their data are also made available for public research both now and in the future, when new analytical techniques may be brought to bear on the specimens. Specimens submitted are required to meet a number of criteria such as having adequate location data and other documentation and be non-toxic and in a dry condition.

Sample submissions vary in size from several hundred to several thousand specimens and may originate from national monitoring programs or local studies, soil science research or ecological studies and may be recently collected or several decades old. Sample use requests are even more varied and range from a single, specific sample to hundreds of samples that are reanalysed with modern laboratory methods. Co-located spectroscopy research by CSIRO has already re-analysed thirty-thousand specimens using NIR and several thousand using MIR [4]. Users of the CSIRO National Soil Archive sometimes only request the soil data, in other cases only the soil specimens. Usually the two go hand in hand. The National Soil Archive aims to balance access to the soil specimens to support current research with the need for preservation of specimens for future generations.

To archive soil specimens, air-dried soil material is transferred into standard long-life containers with bar-coded labels. Associated soil site description, morphology and chemistry data is transcribed and loaded into a standardised relational database (NatSoil) [5]. The site locations and data on soil morphology and chemistry can be accessed via the Australian Soil Resource Information System (ASRIS) [6] and through the SoilMapp for iPad app.

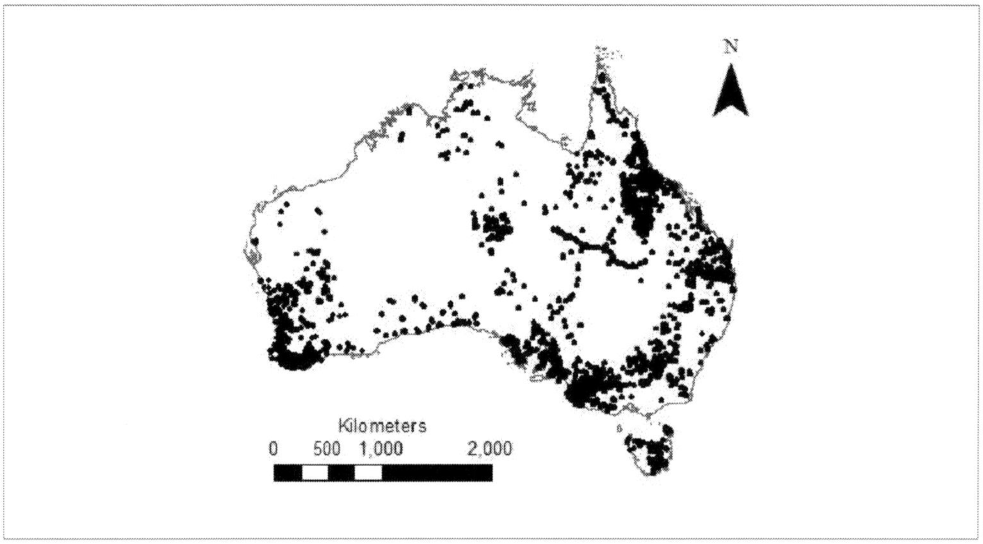

Figure 1. Soil profile locations of archived soil specimens in the National Soil Archive.

3. Who contributes to the Collection?

The national collection was started in 2003 by combining all roughly 24,000 specimens from the former CSIRO Divisions of Soils campuses. These specimens are very rich in data, since they were collected specifically for soil research. This part of the collection reflects the changes in research focus over the last 50 years. Older specimens stored by the CSIRO National Soil Archive (from the 1920's and the 1930's) do not have location data of the currently required accuracy and are therefore not archivable, but that part of the collection can still be used, if one is familiar with the publications. These specimens were collected prior to widespread application of pesticides and herbicides and before the start of nuclear testing in 1945, so provide an important baseline of soil condition. The very old specimens and other yet unarchived specimens are referred to as 'stored' specimens.

Large projects such as the Terrestrial Ecosystem Research Network (TERN) which connects ecosystem scientists and enables them to collect, contribute, store, share and integrate data across disciplines have contributed their AusPlots specimens for safe storage and future use [7]. Other national projects such as the Soil Carbon Research Program (SCaRP), which provides a nationally consistent assessment of soil carbon condition across the major land-use/soil type combinations used for agricultural production, have also contributed specimens [8]. Early discussions between the National Soil Archive and these large projects, which may involve tens of thousands of soil specimens, have led to streamlining of the archiving process.

The National Soil Archive holds the data and soil samples that have been captured since 2004 for the Soil Condition Evaluation and Monitoring (SCEAM) Project, which provides benchmark data for a range of soil condition indicators. For this study all major land-uses in Tasmania - intensive cropping, grazing, native and plantation forestry – are targeted and re-sampling occurs at five year intervals [9].

Some examples of retrospective archiving of specimens from the 1990's are provided by the submission of 1200 well-documented soil specimens from the Desert Uplands Strategic Land Resource Assessment (DUSLARA) in north and central western Queensland and a set of over 4,000 soil specimens with complete data from the Western Australian Department of Environment and Conservation whose dozens of ecological studies focus on remote, non-agricultural, remote areas.

4. How is the Collection used?

The National Soil Archive supports many soil research projects, for example –

- A rapid assessment of the distribution of soil with toxic levels of boron, which was completed by analysing existing specimens at half the estimated expense of new field sampling
- Thousands of archived specimens have been scanned and the MIR and vis -NIR spectra obtained. Using existing limited laboratory data and these spectra, models have been constructed and important soil properties predicted
- The Australian Federal Police have used archived soil specimens to examine the potential of new forensic methods
- A study on acid sulfate soils re-analysed archived specimens from the 1920s to compare to current soil conditions in particular locations
- The Victorian DEPI has re-analysed specimens from the 1970s National Soil Fertility program to study changes in soil carbon content since that time

5. An opportunistic Archive

The fact that the specimens in the National Soil Archive derive from a host of projects, agencies and researchers over an 80 year period adds to the breadth of the collection, but it can also cause some difficulties for users. In terms of data, older collections may be geo-referenced post-collection, which may introduce location errors. The data associated with specimens varies widely in its level of detail provided, for example, the precision of the geographic coordinates of the collection location, morphology description, collection method and analytical methods used.

During archiving, legacy soil data needs to be transcribed from paper copies of internal reports, ASCII text, Fortran based non-relational flat file database files or spreadsheets to fit the dedicated relational database. This involves a substantial effort which requires professional staff because the re-interpretation of descriptive codes, recalculation of measurement units, methods and nomenclature requires familiarity with current and previous classification systems.

For example, when Victorian researchers used archived soil data from the 1970's, some serious implications for using legacy data when dealing with Soil Organic Carbon came to light. A shift in Organic Carbon results appeared due to lab-based modifications over a 40 year period of the Walkley-Black method used to determine OC (Robinson, pers. comm.). The researchers decided to re-analyse archived specimens to develop a correction factor, something that has been done before by [10] in the Carbon conversion factors for an historical data project.

6. Partners

The maintenance of soil archives is not trivial and requires continued financial support and institutional commitment [11]. The National Soil Archive is one of the key deliverables of the Australian Collaborative Land Evaluation Program (ACLEP). ACLEP is a partnership between CSIRO, the Australian Government's Department of Agriculture and the state and territory agencies responsible for land resource assessment.

7. Conclusion

The collection of soil specimens in the CSIRO National Soil Archive is an irreplaceable and valuable resource for current future research due to the number of specimens, the quality of the information available for them and the broad range of sampling locations. Archived soil specimens can support problem-solving soil research by allowing calibration of new analytical methods, enabling the development of baseline national soil property maps and by providing soil material for carbon cycle research.

Care and commitment are needed to establish a world-quality soil archive and the long-term value ought never to be compromised by short-term gains. Decisions of policy makers of the future must rest on the best possible data [12]. To secure the future of the National Soil Archive it is essential that its importance is recognized by the whole scientific community and, importantly, by those who fund the work. This requires foresight, patience and an awareness of the great potential value of the information that can emerge from archived samples. It also involves collections owners taking responsibility for their collections and being held accountable for their long-term care. Besides being an unrivalled resource for addressing many established issues, the National Soil Archive has the potential to provide answers to agricultural and environmental problems not yet recognised.

References

[1] Rothamsted Research 2006 Rothamsted Research: Guide to the Classical and other Long-term Experiments, Datasets and Sample Archive.

[2] Likens GE, Bormann FH 1995 *Biogeochemistry of a Forested Ecosystem.* Second Edition, New York: Springer-Verlag.

[3] Merry, RH 1984 A summary listing of the archival soils collection of the Division of Soils. CSIRO Division of Soil Divisional Report 80, Adelaide.

[4] Viscarra Rossel RA, Walvoort DJJ, McBratney AB, Janik LJ, Skjemstad JO 2006 Visible, near infrared, mid infrared or combined diffuse reflectance spectroscopy for simultaneous assessment of various soil properties. *Geoderma* **1131**, 59-75.

[5] Jacquier DW, Wilson P, Griffin T and Brough D 2012 Soil Information Transfer and Evaluation System (SITES) – Database design and exchange protocols Version 2.0. CSIRO Land and Water report, Canberra.

[6] McKenzie NJ, Jacquier DW, Maschmedt DJ, Griffin EA, Brough DM 2012 The Australian Soil Resource Information System (ASRIS) Technical Specifications. Revised Version 1.6, June 2012. The Australian Collaborative Land Evaluation Program. CSIRO Land and Water report, Canberra.

[7] White A, Sparrow B, Leitch E, Foulkes J, Flitton R, Caddy-Retalic S 2012 AusPlots Rangelands Survey Protocols Manual Version 1.2.9. University of Adelaide Press. Adelaide.

[8] Sanderman J, Baldock J, Hawke B, Macdonald L, Massis-Puccini A., Szarvas S 2011 National Soil Carbon Research Programme: Field and Laboratory Methodologies. CSIRO Land and Water Report, Adelaide.

[9] Cotching W, Kidd D 2010 Evaluation of surface soil condition in Tasmania, Australia. 19th World Congress of Soil Science, Soil Solutions for a Changing World 1-6 August 2010, Brisbane, Australia. Published on DVD.

[10] Skjemstad, JO, Spouncer, LR and Beech, TA 2000 Carbon conversion factors for historical soil carbon data. National Carbon Accounting System Technical Report No. 15, Australian Greenhouse Office, Canberra.

[11] Boone RD, Grigal DF, Sollins O, Ahrens RJ, Armstrong DE 1999 Soil sampling, preparation, archiving and quality control. In *Standard soil methods for long-term ecological research.* In 'Long-term ecological research network series No. 2' (Eds GP Robertson, DC Coleman, CS Bledsoe, P Sollins) pp. 3-28 (Oxford University Press: New York).

[12] Leigh RA, Prew RD, Johnston AE 1994 The management of long-term agricultural field experiments: procedures and policies evolved from the Rothamsted classical experiments. In' *Long-term experiments in agricultural and ecological sciences*'. (Eds RA Leigh, AE Johnston) pp. 253-268 (CAB International: Wallingford).

IOP Conf. Series: Earth and Environmental Science **25** (2015) 012022 doi:10.1088/1755-1315/25/1/012022

Accessing Queensland's soil information – an open data revolution!

Kelly Bryant[1], Lauren O'Brien, Daniel Brough
[1] Queensland Department of Science, Information Technology, Innovation and the Arts, Dutton Park Qld, Australia.

E-mail: Kelly.Bryant@science.dsitia.qld.gov.au

Abstract. The Queensland government is the custodian of soil and land resource information with an estimated value of $75 million. The Soil and Land Information (SALI) system houses this data from over 600 distinct studies with some 96,000 soil observations dating back to the 1940s. This data is now not only used by government but by universities, councils, landowners, consultants and schools. Providing this information to the public in an easy and accessible way, with a focus towards online delivery is crucial. Previous issues with distribution of online soils information in Queensland have stemmed not only from limits to technology but also, changing departmental structures and multiple websites. The department which manages soils information in Queensland has undergone nine name changes in the last 12 years due to Machinery of Government (MoG) restructures. This constantly changing web presence and branding is as confusing for people sourcing soils information as it is for those providing it.

The Queensland government has now moved to a whole of government online environment. This is a single website with no reference to the convoluted structures within government or department names. The aim is to prevent impacts from future MoG changes on the provision of data and information to the public. Queensland government soils now has a single dedicated website (qld.gov.au/environment/land/soil) which has allowed us to start to build a repository for soils information and is a single portal for people to access soils data. It has been demonstrated that this consistent approach to websites improves trust and confidence of users [1] and from this, confidence in using Queensland soils information and data and ultimately better land management decisions.

1. Historical issues with accessing soils information in Queensland

Where do I get soils information in Queensland?

This question would have been an easy answer up until the early 2000s. The Department of Primary Industries (DPI) and the Department of Natural Resources (DNR) were around collectively for 38 years (33 years for DPI and 5 years for DNR). These departments became common brands for the place to go to get soil and land information in Queensland. In the past 12 years the Queensland state department which delivers soil and land information has had seven names (Figure 1). This makes it difficult to become familiar with who to contact for information. Currently to access soils information the contact is through the Department of Science, Information Technology, Innovation and the Arts. To most people this would not be a first port of call.

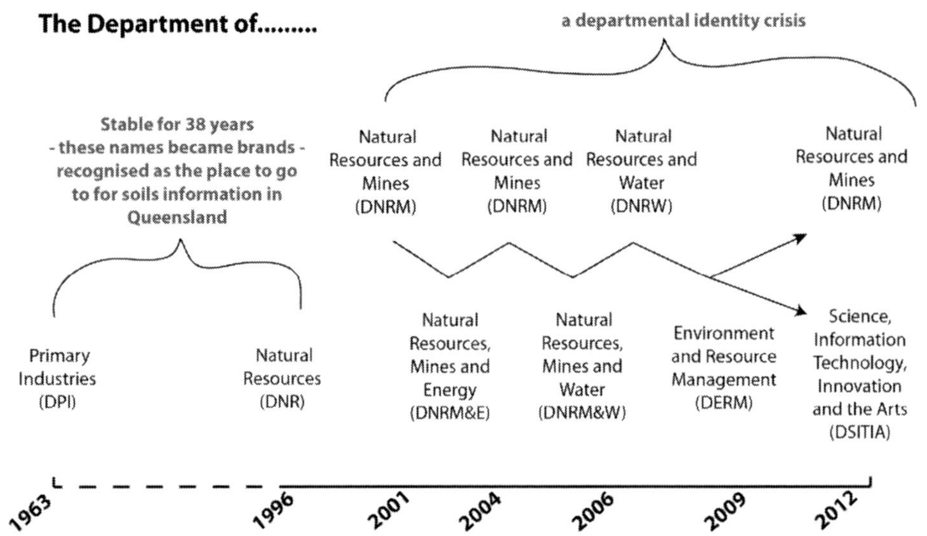

Figure 1. Queensland government's departmental name changes 1963 to 2012.

2. Whole of government sites and stable URLs for delivering information

The Queensland government has moved to a whole of government or "one stop shop" website structure. All Queensland government websites will have the same look and feel and the entry to the site is through a single URL www.qld.gov.au. There were two main aims to achieve by moving to this structure:

1) To be client focused—ensuring that information online is provided based on what the public want. This information comes from focus group research to determine the type of information that people want to access on a state government website site relating to the environment. Previously information provided was governed by each department.

2) To be content focused—focusing on the information provided rather than the department that provides it. Information may be produced by several departments i.e. soils information is currently provided by two different departments, but this is not apparent on the website. This ensures that future restructures of departments do not impact on information delivered through state government websites.

3. Who uses information and what do we provide?

Now there is a stable environment for us to build a repository of soils information the focus has shifted to determining who are our audience is, what information they need and how can we provide this.
The users vary considerably - from people wanting to know what soil is good for growing veggies in their backyard to statewide energy providers planning major infrastructure developments. Over the past three years we have been monitoring the requests that come via phone and email to capture the type of information people are requesting. The majority of requests are for soils mapping and data. In 2012 the Queensland government introduced an open data program (data.qld.gov.au) with the aim of making all data collected by the Queensland government available to the public. Since December 2012 the majority of Queensland soils mapping has been available through this download service – now over 120 soil surveys are available. Currently about nine data sets are downloaded per day through this service, compared to previous requests (either by phone or email) of about nine data sets per month.

4. Future ways to target

Generating statistics on users and information accessed is the best way to target the areas that we need to improve. We have tracked the broad groups of people who use this data to date (Figure 2). The use of more advanced services (such as Google Analytics) in the future will provide a greater level of detail on user groups and information accessed.These types of statistics are essential to enhance information delivery services as they improve our understanding and targeting of user groups.

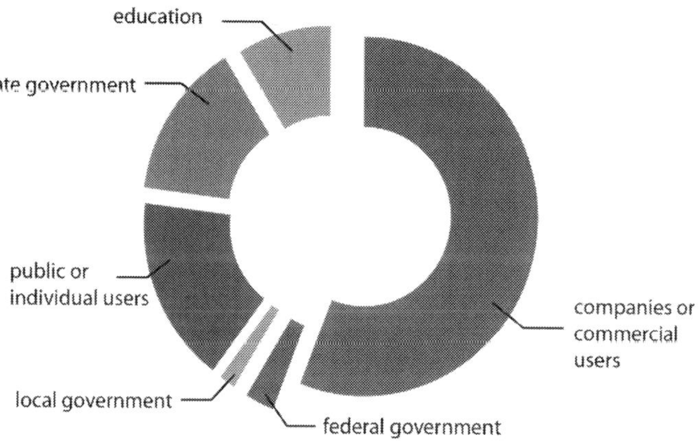

Figure 2 Proportion of soil data downloaded by user groups

References

[1] Queensland Government 2011 Consistent User Experience 3.0, viewed 25/10/2013, http://www.qld.gov.au/web/cue/standard/

The rise of information science: a changing landscape for soil science

Pierre Roudier[1], Alistair Ritchie[2], Carolyn Hedley[1], David Medyckyj-Scott[1]

[1] Landcare Research, Manaaki Whenua, Private Bag 11052, Manawatu Mail Centre, Palmerston North 4442, New Zealand
[2] Landcare Research, Manaaki Whenua, PO Box 69040, Lincoln 7640, New Zealand

E-mail: roudierp@landcareresearch.co.nz

Abstract. The last 15 years have seen the rapid development of a wide range of information technologies. Those developments have been impacting all fields of science, at every step of the scientific method: data collection, data analysis, inference, science communication and outreach. The rate at which data is being generated is increasing exponentially, giving opportunities to improve our understanding of soils. Parallel developments in computing hardware and methods, such as machine learning, open ways to not only harness the ``"data deluge", but also offer a new way to generate knowledge. Finally, emerging data and information delivery protocols are leveraging the outreach power of the World Wide Web to disseminate scientific data and information, and increase their use and understanding outside the boundaries of a given scientific field. However, the nature of this data is mostly new to soil science, and requires adaptation to its diversity and volume. In particular, the integration of the significant amount of legacy soil data collected throughout decades of soil science can be problematic when all necessary metadata is not available. Likewise, knowledge accumulated by our scientific field needs to be acknowledged by - rather than opposed to - numerical methods. While the introduction of this set of emerging technologies is enabling soil science from different points of view, its successful implementation depends on the ability of soil scientists to act as knowledge brokers and support numerical methods.

1. Introduction

The explosion of information technologies has been the most important development in science over the last 15 years. In particular, it has permitted data to be generated at higher rates than ever before and keeps on increasing exponentially. This phenomenon has been described as the "data deluge"'. It has had a considerable impact on science in general: whole disciplines have moved from data-poor to data-rich in a significant shift from a situation where the collection of one data unit is more costly than its analysis, to the opposite situation. Soil science is no exception. Further, as technologies underlying the data deluge develop and mature, they will continue to change the way soils are studied.

Simultaneously, analysis capabilities have been growing to support the exponential increase of data being created [1]. Computing hardware underwent drastic developments in the last decade enabling the analysis of the increased flow of data. Key technology advances have dramatically lowered the cost of storage, amplified by the use of commodity hardware, enabling modern, workload-driven storage of data that is flexible and easy to scale. The invention of high-performance computing (HPC), servers linked up to behave as one, and parallel programming facilitates the use of high numbers of computing cores to solve complex problems [2]. Finally, overarching the increased amount of available data and the development of high-performance computing hardware, the research field of artificial intelligence has produced different numerical methods that are able to handle the large and varied datasets becoming available. Amongst these, machine learning are prediction methods that induce new properties from existing data. Related to machine learning methods, data mining are methods to discover new knowledge from exploring relations in the existing data.

The implications of these technological developments for science are truly disruptive. Pioneering computer scientist Jim Gray has been formulating this as a new paradigm for science exploration and discovery: the fourth paradigm. According to this new scientific paradigm, technologies would lead to the collection, analysis and visualisation of increasingly large amounts of data that would change the very nature of the scientific discovery process [3]:

> The world of science has changed, and there is no question about this. The new model is for the data to be captured by instruments or generated by simulations before being processed by software and for the resulting information or knowledge to be stored in computers. Scientists only get to look at their data fairly late in this pipeline. The techniques and technologies for such data-intensive science are so different that it is worth distinguishing data-intensive science from computational science as a new, fourth paradigm for scientific exploration.

Physicists, astronomers, biologists, chemists, earth and social scientists are all benefiting from access to the tools and technologies that integrate "big data" technologies into standard scientific methods and processes. This paper will discuss both the opportunities and the risks that are associated with the growing influence of these technologies in soil science.

2. The impact of big data technology developments in soil science

2.1. The data deluge in soil science

The "data deluge" is a term coined to describe the acceleration of the production of data over the last decade. The structural changes brought about by the data deluge have originally been described using the three "Vs" [4]:

Volume: the amount of data generated is very important, and challenges the way data has been processed using traditional approaches. [2] reported that the world's information storage capacity grew at an annual rate of 25% between 1986 and 2007 to support the acceleration of data generation, but the amount of information created exceeds the world's storage capacity [5, 1],

Variety: the sources of scientific data, and the form into which it is formatted, are increasingly variable and diverse. The collected data can either be structured or unstructured. This requires tools that are able both to cope with increasingly multi-dimensional problems, and to interact with different sources of data and combine these into a useful set for further analysis,

Velocity: the speed at which data is being created is faster than ever before, due in particular to the increased availability of various sensors. The ever-increasing flow of data requires systems that can support updates on the data on which its analysis rely.

A fourth "V" has been added since this original formulation:

Veracity: refers to the problems of data quality, and to the various levels of uncertainty associated with each of the datasets used to support a decision. Data adequacy, a concept integrating the uncertainty of the dataset and its relative relevance to answer a given question, needs to be assessed on a case-by-case basis.

The terms most widely used to describe the evolution of data ("big data", "data deluge") are unfortunate: what characterises the emerging kinds of data is more than just its volume, the various forms it takes or the rate at which this data is streaming. What makes it really new is its connectedness, and the fact that the value associated with these datasets is substantially increased when combined [6]. The four "Vs" are an opportunity for new insights into soil science: fostered collaborations, scaling-up of scientific concepts by patching together dispersed datasets, meta-analysis of previously published works, discovery and modelling causality in an ensemble of complex datasets, real time analytics and processing.

All these aspects represent a significant challenge to soil science. Traditionally, soil data, because it is generated from field observations and further lab analysis, is scarce and costly [7]. These measurements are very demanding, both in time and associated costs, which is why the number of locations sampled is usually low. But the development of sensing methods proposes an alternative type of soil data: soil properties can be observed indirectly, but at a much higher rate, both in time and space, and for a fraction of the price per sample.

An important part of the data deluge is generated by sensing environmental conditions. An example of such developments is proximal soil sensing (PSS). Proximal soil sensors enable field analysis of soil attributes, such as soil carbon and moisture, and on-the-go sensor surveys can define the spatial variability of these soil attributes. Further, wireless sensor networks can be deployed to monitor the temporal change of the spatially defined soil attributes. These tools make it possible to map and monitor changes of soil attributes in the soil profile at a resolution, both in space and time, that was previously impractical. They also enable a scientist to measure those soil properties in situ, with limited disruption of the soil. A review of the ever-growing array of sensors available for soil applications is provided by [8]. Another example of developing data available for soil scientists to use is remotely sensed environmental data such as LiDAR for high-resolution elevation models, or satellite missions such as Landsat or ASTER [9].

The form that the flow of incoming data takes is very different from the traditional soil datasets, and requires the scientist to adapt and make more robust the traditional data management methods. The use of emerging data in soil science comes with a trade-off: the on-the-spot accuracy of models using these data sources is generally lower than the traditional lab measurements, but is able to capture a greater part of the important spatial variability observed in soils because more locations can be sampled.

2.2. *Development of analysis capabilities to address data growth*

The technological ability to process data has been growing in parallel with data volume and velocity [2]. This rapid development has been two-fold: computing hardware on one hand with high-performance computing (HPC) platforms making use of a number of cores; and machine learning methods on the other hand making the most of the HPC capability and providing mathematical tools that can investigate complex systems using quantitative methods and extract information from all this data. This parallel development fostered a new paradigm for science: using data mining techniques, scientists can try to discover new patterns from the data itself, rather than formulating a hypothesis and testing it empirically [10]. Such approaches have been successfully applied in soil science, and pedotransfer functions (PTF) are generated by interrogating important soil profile databases [11].

Environmental data, with satellite missions such as Landsat or SRTM, are providing soil scientists with environmental data at a very good spatial and temporal resolution. As a consequence, scientists are provided with prior information on the areas they may wish to study, and such environmental covariates can be used to inform their sampling design [12]: for example, the sampling locations at which soil spectra should be taken, or at which the soil moisture sensors should be installed, could be determined by interpretation of a stack of environmental data layers.

Digital soil mapping [13] is another application of machine learning methods in soil science: soil profile data, whether generated using traditional lab methods or using sensing methods such as visible near-infrared spectroscopy, can be combined with environmental covariates to predict soil attributes over the landscape [14], or to improve the spatial resolution of existing data [15]. On top of generating new (spatial) information, data mining has also proved to be a useful tool for knowledge discovery from soil maps derived by soil surveyors [16], or from analyzing large soil databases [17]. In this instance, these methods provide tools to investigate the very complex interactions between soil properties, which are difficult to measure, and a range of widely available environmental covariates derived from proximal or remote sensing technologies.

The application of quantitative methods to soil science, either to predict new soil properties from measured ones (PTF) or to map given soil properties in space (DSM), is not new and has been occurring in our field for the last two decades at least. However, the development of the computing power necessary to process the ever-increasing amounts of available environmental data is occurring on alternative platforms for soil scientists. As computing power is developing, the amount of data to be processed has kept up well with Moore's law. As a consequence, while most of the computing tasks have traditionally been done on single core desktop computers, computing efficiently now requires to (i) move to high-performance, multi-core cluster computers, and (ii) to do the processing task close to the data to limit the input/output overheads associated with moving big chunks of data around. To enjoy the benefits of high-performance computing, soil scientists need to adapt their analysis and tools, and change the way they currently store and distribute data.

3. Challenges and opportunities

3.1. Keeping research reproducible

The concept of reproducibility is at the very heart of the scientific method [18]. The ability for scientific results to be reproduced by peers is the main method to confirm the validity of those results. To achieve this, data and methods need to be accessible to peers. Moreover the scientific method is based on an incremental process: new results generally build on top of previous results, so it is important that researchers have the ability to access and use the methods developed to obtain the state-of-the-art results to generate new science. On top of guaranteeing the necessary reproducibility of the claimed results, data and model availability are working together to accelerate scientific progress and upscale [19].

Because the technical implementation of modern machine learning tools is impacting on the analysis they are supporting, their use is a challenge to research reproducibility. Beyond the parametrisation of the method, there are different ways to implement a given machine learning algorithm [20]. In the past, both data and methods have been reported in scientific papers. The rise of numerical approaches based on various and voluminous data sources and running on HPC platforms is challenging the way science reproducibility has been ensured traditionally. In its current format, the "material and methods" section of scientific papers does not allow for reproducibility, because data has grown much too large to be included, and because the implementation of modern machine learning tools is impacting on the analysis they are supporting [21]. To ensure reproducibility and allow the fostering of science by sharing not only algorithms but also their software implementation, several authors are making the case for sharing code in addition to data.

3.2. Open soil science

In line with these developments, empirical data suggests there is a significant trend in scientific journals to make not only data, but also code available alongside papers published in their columns [20, 22, 23]. Beyond reproducibility, making data and code available are a way to accelerate the pace of research, and improve its cost-efficiency [24]. This idea is formalised by the concept of open science, which promotes open access of science articles along with the availability of code and data [18]. Open science is a virtuous circle where available data can improve models, available models can improve data collection, and open access publication spread ideas further and faster, thus improving the returns on the science endeavour.

Solutions have been proposed. Coming from the software development community, public code repositories like Github now support this evolution giving the opportunity to associate a digital object identifier (DOI) to a code repository [25]. A DOI is a way to uniquely identify an electronic object and its metadata, allowing it to be formally cited in a journal article, accessed by other researchers and recognised as a piece of research. For data, on-line repositories such as Figshare and Zenodo also allow the publication of any research outputs, such as datasets, as a citeable object in its own right.

3.3. Integrating past, present and future data

By their nature, data-oriented approaches often weave various sources of data together. Data availability and structure are guiding the pace at which such science can progress, which in turn raises the value of datasets that have been collected in the past. The emerging sensing techniques, for example, are not exclusive to traditional approaches: reference data is indeed needed to calibrate most quantitative, sensor-based approaches. Legacy lab data can play this role, but its harmonisation and integration with sensing data can be problematic depending on the state of the associated metadata. Unfortunately, scientific data has traditionally been managed on a per-project basis, making data re-use a very challenging task. A consequence of this is "dark data" – data that has been collected but has become invisible to scientists because it is inaccessible [26]. Shining a light on "dark data" by making it available will ensure it can be used for further applications, improving the cost-efficiency of such datasets. It is an opportunity for soil science to give a second youth to the significant amount of legacy data that has been collected over time by soil scientists [27].

The rescue of soil science's data legacy has been one of the major drivers of the GlobalSoilMap project [28]. But while making legacy datasets available in a digital form is necessary for them to be distributed and used in coordination with other existing data, it is not enough, legacy data needs to have metadata associated to it, and its quality needs to be assessed and raised to modern standards [27]. Additionally, when collating several soil datasets, it is also necessary to make them compatible using harmonisation techniques [29, 30]. Numerical techniques are being developed to do so [31], but the use of semantic mediation techniques, such as data mapping and linking based on common information concepts and controlled content (e.g. ontologies, taxonomies, glossaries, linked data, etc.) needs to be investigated in order to facilitate data integration more systematically.

3.4. From sensors to analysis: using interoperability specifications

The drive towards the fourth paradigm requires researchers to make their data publicly available, but also to improve the curation of past datasets to improve their cost-efficiency and allow new results to be derived even after the project is finished. Data-rich approaches underline the need to extend the life cycle of the datasets, and to foster their re-use. While data curation is a new task to add to the soil research departments, those technologies also provide solutions to share and distribute data and methods used in science.

Making data available online is a challenging task. Data needs to be presented in a standard form, and associated with its metadata, so it can be linked with other existing datasets to be truly useful [32]. A set of tools to do so is provided by the data modeling methods, which were first introduced in the domain of software development. Data modeling methods help formalise the scientific concepts and objects that scientists manipulate, along with their relationships. These data models explicitly define the structure of data. In soil science, the ANZSoilML initiative [33] aims to model soil science data from the pedologists perspective. The information model describes and organises soil data, and also provides for the capture of necessary metadata, ensuring the data generation process is documented, and reproducible. Data models like ANZSoilML, if adopted by the community, also ensure the interoperability of the various data sources, which lowers the cost of the data collation for the scientist.

Such data models can then be implemented in an operational solution using web services. These are standardised protocols to stream data over the Internet rather than store it in downloadable files. The advantages of web services over static files are that they allow data to be more dynamic, providing users with updates as data change. Web services also allow the use of new tools, beyond the traditional spreadsheet on a desktop computer, facilitating the weaving of various data sources on various computing platforms (desktop, cloud, but also mobile). To facilitate data sharing, the use of existing standards (such as the Open Geospatial Consortium web services), when appropriate, is encouraged [34].

3.5. Adapting the analysis workflow

In order to keep up with the increasing amount of data generated, HPC and machine learning methods have been very active fields of research -- even going beyond expectations by proposing a fourth paradigm for scientific discovery based on data [3, 10]. The trade-off from these new tools is that researchers need to embrace specific tools to benefit from the technological breakthroughs they provide. An example of such trade-off is that HPC platforms do not generally come with a graphical user interface: the analysis is sent to the computing cluster as a script (i.e. a high-level programming language that is interpreted by another program at runtime rather than compiled by the computer's processor as other programming languages such as C or C++).

While the learning curve associated with these tools might be steep for some, the input of soil scientists into this process is crucial. First, domain knowledge is necessary to ensure sensible use of those technologies, i.e. that they not only provide results that are compatible with the existing knowledge base in soil science, but they also answer valid questions. Moreover, the end goal of science is to produce knowledge, as opposed merely to turn data into information. The role of the domain expert to this regard is two-fold. It is to aggregate the information patterns detected by machine learning methods with the existing knowledge base so they can be turned into knowledge itself and be used to build on; it is also to ensure that the staggering amount of knowledge (i.e. largely undocumented experience, insights, knowledge and skills) accumulated throughout decades of soil science in the lab and in the field is acknowledged and used. Numerical data can benefit from it and support the generation of new knowledge, but not replace the soil scientist's domain expertise. In this sense, the input of soil scientists as knowledge brokers for the successful implementation of data-oriented methods.

3.6. Opportunities in soil science outreach

The contribution of information technologies to the science endeavour is not limited to data generation and analysis; innovations have also taken place in how to share and present data. The emergence of the World Wide Web as the main communication platform and the development of mobile technologies are examples of this. Developments in web technologies allow advanced software applications to be used in many different computing environments as opposed to traditional desktop-side technologies (an example of this is web mapping technologies such as Google Maps, which have popularised maps as a form of information presentation, previously limited to user-unfriendly desktop GIS environments). Web data services, such as OGC web services, offer a standardised way to "stream" data over the web. Those services can connect a data provider to a range of data consumers, inside or outside the research community, but can also connect the data provider directly to processing platforms. The agility of data services technologies can improve soil data dissemination, and improve the return on investment of soil data but maximising data re-use. Moving data provision on the server side does however make it vulnerable to problems associated with cloud storage, such as unexpected server downtime. It also raises synchronisation needs for offline applications, a common case when working in the field. These technologies are changing the way science is shared across scientific communities and delivered to policy makers and how science interacts with the general public, by providing new tools and the communication channels to do so. In particular, mobile devices and data visualisation techniques can build upon those data streams, and tailor the way science can be consumed by a variety of stakeholders [35].

Information technologies are also fostering and enabling feedback loops from the stakeholders to the researchers. Analysing the logs of web sites and web services can assess whether scientists are or are not answering the questions asked by the public. Additionally, most smartphones carry a GPS chip, a colour high-resolution CCD camera and full-time internet connection, can be used as sensors [36], and even contribute to the data collection effort [37] or validation of data.

4. Conclusion

Disruptive developments in the field of information technologies impact on all aspects of science, and soil science is no exception. With the contribution of technologies such as remote sensing, proximal soil sensing and wireless sensor networks, the data used to improve our understanding of soil is becoming more intensive and more dynamic. It is also more diverse, and while the number of sources of soil data is increasing, it is important to combine these to create potentially useful soil information. The integration of this emerging stream of data with existing soil data is also important in order to leverage and acknowledge the data collected over the last decades using traditional approaches. The data management, in particular the available metadata of these legacy datasets, will have an impact on the feasibility of this integration. High-performance computing hardware, machine learning and data mining methods are supporting these important changes in data, which allow us to maximise the value of our data. The combination of data richness and HPC hardware enables numerical approaches in soil science, previously impractical, based on the recent developments in machine learning methods.

However, it is important to notice that the challenges raised by the "data deluge" are not so much the amount of data per se, but the novel analysis to produce knowledge and help insightful decisions [38, 1]. While the amount of environmental data has increased considerably over the last decade - so much that soil scientists could become overwhelmed by the "data deluge", the "data revolution" represents a positive challenge for soil science. But to achieve this goal, it is crucial for soil scientists to be involved alongside data and computer scientists to ensure that expert knowledge is captured and used to validate results, and to acknowledge and preserve legacy data.

The fourth paradigm is a new way of generating scientific knowledge building on the "data deluge" and on the "big data" technologies. Associated upcoming changes require scientists to adapt their tools and re-skill themselves to fully benefit from these developments. In particular, a shift towards data-driven approaches requires data to be openly available, support reproducibility of the scientific results, and foster the upscale of those results. Ensuring the interoperability and availability of soil data is not only a requirement for the emerging sensing technologies, but can also improve the visibility of the staggering amount of soil information that is currently inaccessible. In particular, setting up soil data services is an import step to improve the dissemination of soil information, and support its use by very various users and across a range of sectors outside of the sole research field of soil science.

References

[1] Hilbert M 2013 Big data for development: from information to knowledge societies *SSRN Electronic Journal* http://ssrn.com/abstract=2205145
[2] Hilbert M and Lopez P 2011 The World's technological capacity to store, communicate, and compute information *Science* **332** 60-65
[3] Gray J 2009 Jim Gray on eScience: a transformed scientific method *The Fourth Paradigm: Data-Intensive Scientific Discovery* ed Hey T, Tansley S and Tolle K (Redmond, WA, USA: Microsoft Research)
[4] Gartner Inc 2011 Gartner says solving 'Big data' challenge involves more than just managing volumes of data http://www.gartner.com/newsroom/id/1731916
[5] Cukier K 2010 Data, data everywhere: a special report on managing information *The Economist* **394** 3-5
[6] Boyd D and Crawford K 2012 *Information, Communication & Society* **15** 662-79
[7] Lagacherie P and McBratney AB 2006 Spatial soil information systems and spatial soil inference systems: Perspectives for digital soil mapping *Developments in Soil Science Volume 31 Digital Soil Mapping An Introductory Perspective* ed P Lagacherie AB McBratney and Voltz M (Elsevier) pp 3-22
[8] Adamchuk VI and Viscarra Rossel RA 2010 Development of on-the-go proximal soil sensor systems Proximal Soil Sensing *Progress in Soil Science* ed Viscarra Rossel R A, McBratney AB and Minasny B (Springer Netherlands) pp 15-28
[9] Mulder VL, de Bruin S, Schaepman ME and Mayr TR 2011 The use of remote sensing in soil

and terrain mapping – A review *Geoderma* **162** 1-19
[10] Hey T, Tansley S and Tolle K 2009 *The Fourth Paradigm: Data-Intensive Scientific Discovery* (Redmond, WA, USA: Microsoft Research)
[11] Bouma J 1989 Using soil survey data for quantitative land evaluation *Advances in Soil Science* (*Advances in Soil Science* no 9) ed Stewart B A (Springer US) pp 177-213
[12] de Gruijter J, Brus DJ, Bierkens MFP and Knotters M 2006 *Sampling for Natural Resource Monitoring* (Springer)
[13] McBratney AB, Mendona Santos ML and Minasny B 2003 On digital soil mapping *Geoderma* **117** 3-52
[14] Behrens T and Scholten T 2006 A comparison of data-mining techniques in predictive soil mapping *Developments in Soil Science Volume 31 Digital Soil Mapping An Introductory Perspective* ed P Lagacherie A B M and Voltz M (Elsevier) pp 353-617
[15] Bui EN and Moran CJ 2001 Disaggregation of polygons of surficial geology and soil maps using spatial modelling and legacy data *Geoderma* **103** 79-94
[16] Bui EN, Loughhead A and Corner R 1999 Extracting soil-landscape rules from previous soil surveys *Aust J Soil Res* **37** 495-508
[17] Bui EN, Henderson BL and Viergever K 2006 Knowledge discovery from models of soil properties developed through data mining *Ecol Modell* **191** 431-46
[18] Ibanez L, Schroeder WJ and Hanwell MD 2014 Practicing open science *Implementing reproducible research* ed Stodden V, Leisch F and Peng RD (Florida: CRC Press) 241-280
[19] Wolkovich EM, Regetz J and O'Connor MI 2012 Advances in global change research require open science by individual researchers *Global Change Biology* **18** 2102-10
[20] Ince DC, Hatton L and Graham-Cumming J 2012 The case for open computer programs *Nature* **482** 485-88
[21] Malik T, Pham Q and Foster I 2014 SOLE: towards descriptive and interactive publications *Implementing reproducible research* ed Stodden V, Leisch F and Peng RD Peng (Florida: CRC Press)
[22] Stodden V, Guo P and Ma Z 2013 *PLoS ONE* **8**
[23] Russel J 2013 Reproducing our irreproducibility *Nature* **496** 398-98
[24] Bloom T 2014 Data access for the open access literature: PLOSs data policy http://www.plos.org/data-access-for-the-open-access-literature-ploss-data-policy/
[25] Smith A 2014 Improving GitHub for science https://github.com/blog/1840-improving-github-for-science
[26] Heidorn PB 2008 Shedding light on the dark data in the long trail of science *Libr Trends* **57** 280-99 ISSN 1559-0682
[27] Rossiter DG 2008 Digital soil mapping as a component of data renewal for areas with sparse soil data infrastructures *Digital Soil Mapping with Limited Data* ed Hartemink AE, McBratney A and Mendona- Santos MdL (Springer Netherlands) pp 69-80
[28] Hempel J, McBratney A, Arrouays D, McKenzie N and Hartemink A 2014 *GlobalSoilMap project history GlobalSoilMap: Basis of the global spatial soil information system* ed Arrouays D, McKenzie N, Hempel J, Richer de Forges A and McBratney AB pp 3-8
[29] Keune H, Murray AB and Benking H 1991 Harmonization of environmental measurement *GeoJournal* **23** 249-55
[30] Arrouays D, Marchant BP, Saby NPA, Meersmans J, Orton TG, Martin MP, Bellamy PH, Lark RM and Kibblewhite M 2012 Generic issues on broad-scale soil monitoring schemes: A review *Pedosphere* **22** 456-469
[31] Baume O, Skien J O, Heuvelink G B M, Pebesma E J and Melles S J 2011 *Inter J App Earth Obs* **13** 409-419
[32] Bizer C, Heath T and Berners-Lee T 2009 *Int J Semant Web Inf* **5** 1-22
[33] Simons B, Wilson P, Ritchie A and Cox S 2013 ANZSoilML: an Australian-New Zealand standard for exchange of soil data *EGU General Assembly Conference Abstracts* **15** 6802

[34] Percivall G 2010 *Int J Digital Earth* **3** 14-30
[35] Müller M, Medyckyj-Scott D, Cowie A, Heuer TH and Roudier P 2013 Current status and future directions of mobile GIS ed Moore A and Whigham PA (University of Otago, Dunedin, New Zealand)
[36] The James Hutton Institute SOCiT URL https://itunes.apple.com/gb/app/socit/id631266307
[37] Goodchild MF 2007 Citizens as sensors: The world of volunteered geography *GeoJournal* **69** 211-221
[38] Jacobs A 2009 *Commun ACM* **52** 36

Institute of Physics
Dirac House, Temple Back
Bristol BS1 6BE UK

CURRAN ASSOCIATES INC
proceedings.com